EARTH SCIENCES IN THE 21ST CENTURY



# ATMOSPHERE AND CLIMATE

# PHYSICS, COMPOSITION/DYNAMICS AND HEALTH IMPACTS

## EARTH SCIENCES IN THE 21ST CENTURY

Additional books in this series can be found on Nova's website under the Series tab.

Additional E-books in this series can be found on Nova's website under the E-book tab.

## CLIMATE CHANGE AND ITS CAUSES, EFFECTS AND PREDICTION

Additional books in this series can be found on Nova's website under the Series tab.

Additional E-books in this series can be found on Nova's website under the E-book tab.

EARTH SCIENCES IN THE 21ST CENTURY

# ATMOSPHERE AND CLIMATE

# PHYSICS, COMPOSITION/DYNAMICS AND HEALTH IMPACTS

ANDREW WRIGHT
AND
SHAWN JOHNSON
EDITORS

*New York*

**Library of Congress Cataloging-in-Publication Data**

ISBN: 978-1-62417-433-9

*Published by Nova Science Publishers, Inc. † New York*

# CONTENTS

# PREFACE

In this book, the authors present current research in the study of the physics, composition/dynamics and health impacts of atmosphere and climate. Topics discussed include global warming in the context of millennial climate history; large ice sheets may have the capacity to self-destruct and cause rapid and permanent changes in global climate and sea level; temporal and spatial patterns of PM10 concentrations in urban areas; and advanced fibrous adsorbents for CO2 removal from power plants and the atmosphere.

Chapter 1 – Recent global warming has greatly reduced the area of perennial sea ice in the Arctic Ocean, caused mountain glaciers to retreat and allowed major ice streams draining the Antarctic and Greenland ice sheets to approach and even exceed surge velocities. This has brought new urgency to the concept that large ice sheets, past, present, and future, may have the capacity to self-destruct causing rapid and permanent changes in global climate and sea level. This possibility has focused attention on these ice streams, beginning with Jakobshavn Isbrae, long the world's fastest ice stream and discharging seven percent of Greenland ice, a discharge rate that has now doubled following catastrophic disintegration of its buttressing ice shelf in Jakobshavn Isfjord in 1997. A geometrical solution to the longitudinal force balance was combined with a simple mass balance to obtain two solutions to the resulting force-and-mass balance from 1986 to 2006. The two solutions are obtained using two approaches to quantify the loss of buttressing. The best match to changes in surface elevations and velocities along the ice stream, especially the rapidly lowering surface during this timespan, is provided by the solution that has the strongest reduction of ice-bed coupling up the ice stream, quantified by a rapid increase in the "floating fraction" of ice. This pattern seems to be repeating for other Greenland and Antarctic ice streams. If rapid surface lowering and ice discharge spreads to all major ice streams and lasts for centuries, even accelerating over time, self-destruction of these two remaining ice sheets seems inevitable. Without the high albedo of ice sheets, global warming would increase until a new global equilibrium with the new boundary conditions was established. The other possibility is the observed rapid ice thinning and discharge may be transient and lead to a new stable equilibrium that keeps these ice sheets largely intact with no significant change in global climate and sea level. Our results from Jakobshavn Isbrae were unable to determine which scenario is most likely, but time will tell, probably over the next few decades as observational data accumulate and our modeling capability improves.

Chapter 2 – Analysis of the available data on the Earth's global temperature over the past 150-1000 years is performed. Both instrumental (ground-based and satellite) data and proxies, obtained by paleoclimatic approaches, are analyzed. The estimation of advantages and shortcomings of various sources of the climatic information is made. Complex research of all the data set has shown that: (a) $20^{th}$ century was actually warm, i.e. the global temperature of the Earth in 1900-2000 was above the millennial average; (b) in Northern hemisphere the time interval 1988-2008 likely was the warmest two decades over the last 1000 years. The potential origins of the considerable growth of temperature are considered. It is shown that the main factors likely responsible for a rise in global temperature are: (a) industrial emissions of greenhouse gases ($CO_2$, methane, nitrogen oxides); (b) natural climatic cycles; (c) solar-cosmic factors, including changes of solar luminosity, variation of the intensity of galactic cosmic ray and cosmic dust flux; (d) regional anthropogenic impact, including cutting down of woods, desertification, irrigation and ploughing of virgin lands. But it is difficult to estimate more or less precisely actual contribution of each factor due to our insufficient knowledge about them. It is shown that despite the successes and hard work in the field of both climatology and paleoclimatology our knowledge about the Earth's past climate is still not fully adequate. That is why it is hard to predict the change of the global climate in the current century. We can only consider some possible scenarios of the future climate evolution: (a) if global warming is a result of only industrial greenhouse effect, the rise of temperature will continue in $21^{th}$ century according to forecasts of IPCC; (b) if global warming is mainly a result of natural climatic cycles, in the first part of $21^{th}$ century the global temperature will decrease.

Chapter 3 – Particulate matter is one of the most important air pollutants that negatively influence human health in Europe. They are emitted by both natural (volcanic eruptions, seismic activity and forest fires) and anthropogenic sources (all types of man-made combustion and some industrial processes). Long exposure to $PM_{10}$ and to $PM_{2.5}$ (particles with an aerodynamic diameter smaller than 10 and 2.5 μm, respectively) has been associated with respiratory and cardiovascular diseases. Air quality is strongly dependent on weather and, therefore, sensitive to climate change. Consequently, temporal and spatial patterns of air pollution are not only dependent on emission changes. The changes in the boundary layer height, relative humidity, temperature, solar radiation, wind speed and precipitation may be responsible for significant differences in air pollutant concentration patterns. Thus, this chapter aims to evaluate temporal and spatial patterns of $PM_{10}$ concentrations at an urban area through the application of principal component analysis (PCA). It was applied to air quality data collected in the Air Quality Monitoring Network of Oporto Metropolitan Area during the period from 2006 to 2010. The monitored region presented high $PM_{10}$ concentrations (with several exceedances to European Union standard limits); however, they presented a decreasing tendency. PCA identified seven to nine $PM_{10}$ pollution patterns. Two monitoring sites often presented similar air pollution pattern. Analysing the annual average profile, $PM_{10}$ concentrations were higher during night time in winter period. This phenomenon is justified by the daily and seasonal variation of atmospheric boundary layer height, which influences the mixing of air pollutants.

Chapter 4 – The mitigation of carbon dioxide ($CO_2$) emissions has attracted worldwide attention because of the environmental and economic threats posed by possible global climate changes. Considering that the utilization of fossil energy carriers will continue and may even increase over the coming decades, advanced $CO_2$ capture materials and technologies are of the utmost importance. Micro/nanofibers have unique porous structure, high surface area, and good mechanical properties and thereby are attractive candidates as advanced sorbents for $CO_2$ removal. Fibrous sorbents may have high $CO_2$ capture capacity and little resistance to bulk gas flow and thereby have fast kinetics. In this chapter, we have summarized the recent progress in the development of fibrous sorbents (e.g. amine-modified fibers, carbon fibers, hollow fiber membranes, etc.) for $CO_2$ removal, described the types of fiber materials that have been studied, and discussed their fabrication variables and $CO_2$ adsorption performance in details. This review may pave the way for developing advanced fibrous adsorbents for $CO_2$ removal from power plants and even from the atmosphere.

In: Atmosphere and Climate
Editors: A. Wright and S. Johnson
ISBN: 978-1-62417-433-9


*Chapter 1*

# CAN ICE SHEETS SELF-DESTRUCT AND CAUSE RAPID CLIMATE CHANGE?

## A CASE STUDY: JAKOBSHAVN ISBRAE, GREENLAND

***Coen Hofstede*[1] *and Terence Hughes*[2]**
[1]Alfred Wegener Institute for Polar and Marine Research, Bremerhaven, Germany
[2]Earth Sciences and Climate Change, University of Maine, Fort Pierre, SD, US

## ABSTRACT

Recent global warming has greatly reduced the area of perennial sea ice in the Arctic Ocean, caused mountain glaciers to retreat and allowed major ice streams draining the Antarctic and Greenland ice sheets to approach and even exceed surge velocities. This has brought new urgency to the concept that large ice sheets, past, present, and future, may have the capacity to self-destruct causing rapid and permanent changes in global climate and sea level. This possibility has focused attention on these ice streams, beginning with Jakobshavn Isbrae, long the world's fastest ice stream and discharging seven percent of Greenland ice, a discharge rate that has now doubled following catastrophic disintegration of its buttressing ice shelf in Jakobshavn Isfjord in 1997. A geometrical solution to the longitudinal force balance was combined with a simple mass balance to obtain two solutions to the resulting force-and-mass balance from 1986 to 2006. The two solutions are obtained using two approaches to quantify the loss of buttressing. The best match to changes in surface elevations and velocities along the ice stream, especially the rapidly lowering surface during this timespan, is provided by the solution that has the strongest reduction of ice-bed coupling up the ice stream, quantified by a rapid increase in the "floating fraction" of ice. This pattern seems to be repeating for other Greenland and Antarctic ice streams. If rapid surface lowering and ice discharge spreads to all major ice streams and lasts for centuries, even accelerating over time, self-destruction of these two remaining ice sheets seems inevitable. Without the high albedo of ice sheets, global warming would increase until a new global equilibrium with the new boundary conditions was established. The other possibility is the observed rapid ice thinning and discharge may be transient and lead to a new stable equilibrium that keeps these ice sheets largely intact with no significant change in global climate and sea level. Our results from Jakobshavn Isbrae were unable to determine which scenario is most likely, but time will tell, probably over the next few decades as observational data accumulate and our modeling capability improves.

## INTRODUCTION

Budd et al. (1971) developed the first computer model that simulated the dynamics of ice sheets and applied it to the Antarctic Ice Sheet. It was a steady-state model applied along surface flowlines normal to ice-elevation contour lines. They used surface accumulation rates, temperatures, and elevations, ice thicknesses, and bed topography to solve the equations of force, mass, and energy balance to obtain internal stresses, temperatures, and velocities, and basal melting rates for prescribed geothermal heat input. The model also computed time horizons at depth for specified surface accumulation rates that could be compared to internal radar-reflecting horizons and oxygen-isotope stratigraphy that were believed to be linked to former surface accumulation patterns along the flowlines. Modest changes in surface accumulation rates and the geothermal heat flux caused a largely frozen bed to become largely thawed with substantial basal melting rates. Over time, this buildup of basal water would decouple the ice sheet from its bed and cause the whole ice sheet to surge. Sugden (1977) corrected this defect by placing a basal freezing zone beyond the basal melting zone, so basal meltwater moving into the freezing zone would freeze onto basal ice and thereby couple the ice sheet to the bed for steady-state subglacial hydrology. He applied his model to the former Laurentide Ice Sheet at the last glacial maximum. These were essentially top-down models that used surface conditions to determine bed conditions.

Mahaffey (1976) made the next major breakthrough by extending flowline solutions to the map plane, so changing flow patterns through time could be simulated. Reeh (1982, 1984) used plasticity theory to reconstruct the Greenland Ice Sheet today and the former Innuitian Ice Sheet in Arctic Canada in the map plane. An early three-dimensional time-dependent model in the map plane was applied to the Antarctic (Huybrechts, 1990, 1992) and Greenland (Huybrechts 1994, 1996) ice sheets. These models produced flow that converged on "ice streams" that coincided with observed ice streams and discharged most of the ice, but ice-stream dynamics were not included in the models. Early models simulated slow sheet flow, which covers about 90 percent of ice sheets, but it became increasingly obvious that up to 90 percent of the ice was discharged by ice streams. Denton and Hughes (1981) made the first attempt to include ice streams in reconstructing ice sheets globally at the Last Glacial Maximum (LGM) 18,000 years ago. They mapped basal thermal zones (melted, melting, freezing, and frozen basal conditions) from glacial geology, and then calculated ice elevations from the strength of ice-bed coupling in these zones. This is essentially a bottom-up approach (Hughes, 1998; 2009a).

Like large rivers that discharge most precipitation over continents, large ice streams are fed by numerous tributaries and discharge most of the ice precipitated over ice sheets (Rignot et al., 2011). Ice streams are fast currents of ice that develop from slow interior sheet flow and deliver internal ice to terminal ice lobes grounded on land or to floating ice tongues that often merge to become floating ice shelves in water. Ice is lost by melting and calving from ice lobes and ice shelves. MacAyeal and Thomas (1982) and Morland (1987) were first to model ice shelves in the map plane. Antarctic ice shelves often lie in confining embayments and are locally grounded where they scrape over basal pinning points to produce surface ice rumples, or flow around pinning points to produce surface ice rises. These boundary conditions allow ice shelves to buttress ice streams that feed them (Thomas, 1973a, 1973b). Unlike sheet flow, where gravitational forcing is balanced against a basal drag force, in shelf flow gravitational forcing is balanced by a tensile force in ice arising from the imbalance in the longitudinal gravitational forces in ice and water, first recognized at the calving front (Weertman, 1957a; Robin, 1958), and later shown to exist from the calving front to the grounding line (Hughes, 2009b).

MacAyeal (1989) made the first attempt to model a transition from sheet flow to stream flow to shelf flow by including basal drag as an addition to stresses resisting gravitational shelf flow. He applied his model to Whillans Ice Stream (formerly ice stream B) that supplies ice to the Ross Ice Shelf in Antarctica. Hughes (1992) introduced a "floating fraction" of ice that decreases from unity at the ice-shelf grounding line to zero at the head of the ice stream. The floating fraction partitioned forces resting gravitational flow in ice streams between basal drag where the ice stream is locally grounded and a tensile force where the ice stream is locally floating.

By the turn of the millennium, Earth-orbiting satellites were mapping substantial reductions of Arctic sea ice, occasionally opening the Northeast and Northwest Passages to the Orient, glacial geologists were mapping retreat of mountain glaciers worldwide, and glaciologists were reporting rapid accelerations of some Greenland and Antarctic ice streams and outlet glaciers, even exceeding "surge" velocities that occur periodically in some mountain glaciers. This was attributed to global warming, and it focused attention of glaciologists on the dynamics of ice streams. A great advance in gathering field data and in developing ice-sheet models followed.

In the field, initial attention was focused on Pine Island Glacier (Rignot, 1998a) and Thwaites Glacier (Rignot, 2001) in Antarctica, and on Jakobshavn Isbrae (Thomas et al., 2003) and on Kangerdlugssuaq Glacier and Helheim Glacier (Stearns and Hamilton, 2007) in Greenland. Hughes (1986) predicted the positive-feedback mechanisms leading to the surge of Jakobshavn Isbrae would move northward along both Greenland coasts as climate warmed, and eventually cause surges of northern Greenland ice streams. He called this "The Jakobshavn Effect" and it has apparently moved to several northern ice streams (Rignot et al., 2001). These positive feedbacks included disintegration of buttressing ice shelves and surface meltwater reaching the bed through crevasses and moulins. Acceleration caused by meltwater reaching the bed has now been confirmed for Jakobshavn Isbrae (Zwally et al., 2002).

Accelerated discharge of ice streams has followed disintegration of buttressing ice shelves, both in Antarctica when the central Larsen Ice Shelf disintegrated (DeAngelis and Skvarka, 2003) and in Greenland when the floating ice tongue of Jakobshavn Isbrae disintegrated (Joughin et al., 2008). Catastrophic ice-shelf disintegration has been linked to warm ocean water passing under Antarctic ice shelves (Jacobs et al., 1996; Shepherd et al., 2004) and entering Greenland fjords (Holland et al., 2008; Rignot and Steffen, 2008). Bottom melting of ice shelves destabilized their grounding lines in both

Antarctica (Rignot, 1998a, 1998b, 2001) and Greenland (Rignot, 1998c; Rignot and Steffen, 2008).

Modeling advances have paralleled field studies. The stability of ice shelves is linked to stress fields that produce crevasses (Rist et al., 1996, 1999, 2002), to catastrophic disintegration of ice shelves along crevasses transverse to ice flow (MacAyeal et al., 2003), to ice dynamics at the grounding line (Hindmarsh and LeMeur, 2001; Schoof, 2007; Schoof and Hindmarsh, 2010) and at the calving front (Kenneally and Hughes, 2006; Hindmarsh, 2012), and to ocean swells that pass under ice shelves (MacAyeal et al., 2009). Modeling basal conditions has advanced from slow sheet flow sliding on a wet bed (Weertman, 1957b; Kamb, 1970) to sliding of fast ice streams that supply ice shelves (Dupont and Alley, 2005a, 2005b). Models now treat the initiation of stream flow (e.g., Hindmarsh, 2009; Hughes, 2009c) and how basal meltwater uncouples ice from the bed beneath ice streams (e.g., Alley, 1989a, 1989b; MacAyeal, 1989; Anandakrishnan and Alley, 1997; Kamb, 2001; Hughes, 2009b; Bueler and Brown, 2009; Bassis, 2010; Schoof and Hindmarsh, 2010, Hughes et al., 2011). Modeling ice-bed decoupling by basal water over time requires modeling subglacial hydrology as basal water flows from sources to sinks (Johnson, 2002; Johnson and Fastook, 2002). In ice streams, the full equilibrium/momentum equations must eventually be solved because all six stresses resist gravitational flow (Sargent, 2009).

## A Holistic Ice-Sheet Model

A holistic ice-sheet model must provide smooth transitions from sheet flow to stream flow to shelf flow in the longitudinal force balance in the direction of gravitational flow of ice. If this force balance is done for flowbands having the width of the ice stream, assumed to be constant, the six resisting stresses reduce to four, a longitudinal tension stress $\sigma_T$ that pulls upslope ice, a longitudinal compression stress $\sigma_C$ that pushes downslope ice, a basal shear stress $\tau_O$ due to basal drag, and side shear stress $\tau_S$ due to side drag. Transverse stresses caused by converging and diverging flow that changes the flowband width can then be ignored. This allows a simple geometrical force balance in the longitudinal direction of ice flow, along which all of these stresses vary with changing floating fraction $\phi$ of ice in the flowband. For sheet flow, $\phi = 0$ when the bed is dry (frozen) and $\phi \to 0$ when the bed is wet (thawed). For stream flow, $0 < \phi < 1$ with $\phi$ generally increasing downstream.

For shelf flow, $\phi = 1$ for a freely-floating ice shelf and $\phi \rightarrow 1$ when a confined and locally pinned ice shelf provides buttressing for the ice stream.

Figure 1 is a cartoon showing places where $\phi = 0$ for grounded ice, wet under an ice stream, and $\phi = 1$ for places where floating ice exists under the ice stream. These places generally correspond to hills and hollows in bedrock topography, or to soft sediments or till that are unsaturated and supersaturated with water, respectively. Bedrock hills and an unsaturated soft bed resist gravitational motion. Taking Cartesian coordinates with x horizontal and positive against ice flow, y horizontal and transverse to ice flow, and z vertical and positive above sea level, at distance x from the ice-shelf grounding line, a flowband of width $w_I$ has floating segments that add up to width $w_F < w_I$ in the ice stream. Floating fraction $\phi$ can then be defined for the horizontal longitudinal force balance as:

$$\phi = \frac{w_F}{w_I} = \frac{h_F}{h_I} = \frac{(\rho_W / \rho_I) h_W}{h_I} = \frac{\rho_W g h_W}{\rho_I g h_I} = \frac{P_W^*}{P_I} \tag{1}$$

where $h_F = h_I (w_F/w_I) = h_I \phi$ is the part of ice thickness $h_I$ supported by basal water, $\rho_W$ is water density, $\rho_I$ is ice density, $P_W{}^* = \rho_W g\, h_W$ is an effective basal water pressure that resists ice flow, $P_I = \rho_I g\, h_I$ is the ice overburden pressure, and $g$ is gravity acceleration. In a vertical force balance, apply Newton's second and third laws of motion to the base of columns having basal area $A_O = w_I\, \Delta x$. Gravity forces $\rho_I\, g\, h_I\, A_O$ and $\rho_W\, g\, h_W\, A_O$ are balanced by pressure forces $P_I\, A_O$ and $P_W\, A_O$, respectively, giving $P_W = \rho_W\, g\, h_W$ as the actual basal water pressure and $P_I = \rho_I\, g\, h_I$ as the basal ice pressure. For ice shelves, $P_W = P_I$ everywhere. For ice streams $P_W \approx P_I$ because basal water flowing from sources to sinks causes variations in $P_W$ that do not coincide everywhere with $P_I$.

Figure 2 shows an exaggerated vertical longitudinal cross-section of a flowband from the ice divide to an ice stream and ending at the calving front of a confined and pinned ice shelf. Flow is from right to left. The top panel shows in shading the part of the flowband that rests on the bed. Solid, broken, and dashed lines show respective heights $h_I$, $h_F$, and $h_W$ above basal ice. The ice shelf lies in a confining embayment grounded along side lengths $L_S$, at an ice rise of circumference $C_R$, and at ice rumples of area $A_R$, so it buttresses the ice stream. Stresses resisting gravitational flow are shown at distance $x$ from the ice-shelf grounding line and averaged over the distance from 0 to $x$. The middle panel shows a large triangular area equal to the gravitational driving force ½ $P_I h_I$.

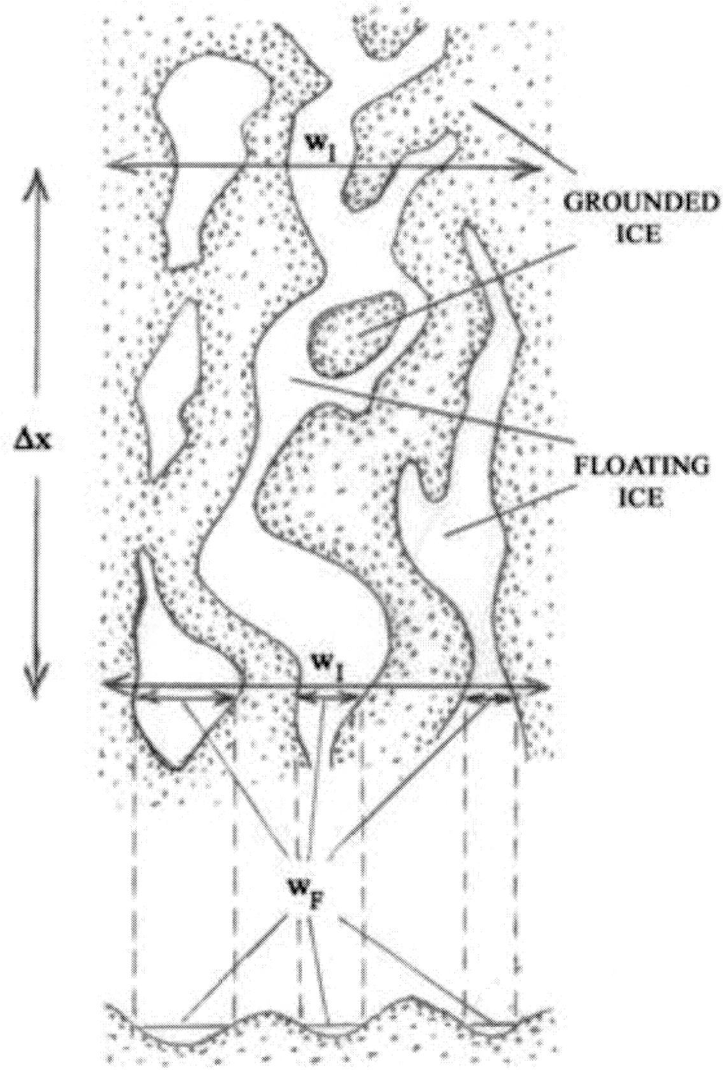


Figure 1. A cartoon of the bed under an ice stream. Ice flow is along incremental length $\Delta x$ in plan view (top) and at $x$ in transverse cross-section (bottom). Ice is either floating above bedrock or supersaturated sediments and till (undotted areas) or grounded on bedrock or unsaturated sediments and till (dotted areas) for respective floating flowband widths $w_F$ and grounded flowband widths $w_I - w_F$. Floating fraction $\phi$ of ice over area $w_I \Delta x$ becomes $\phi = w_F / w_I$ at $x$ when $\Delta x \to 0$.

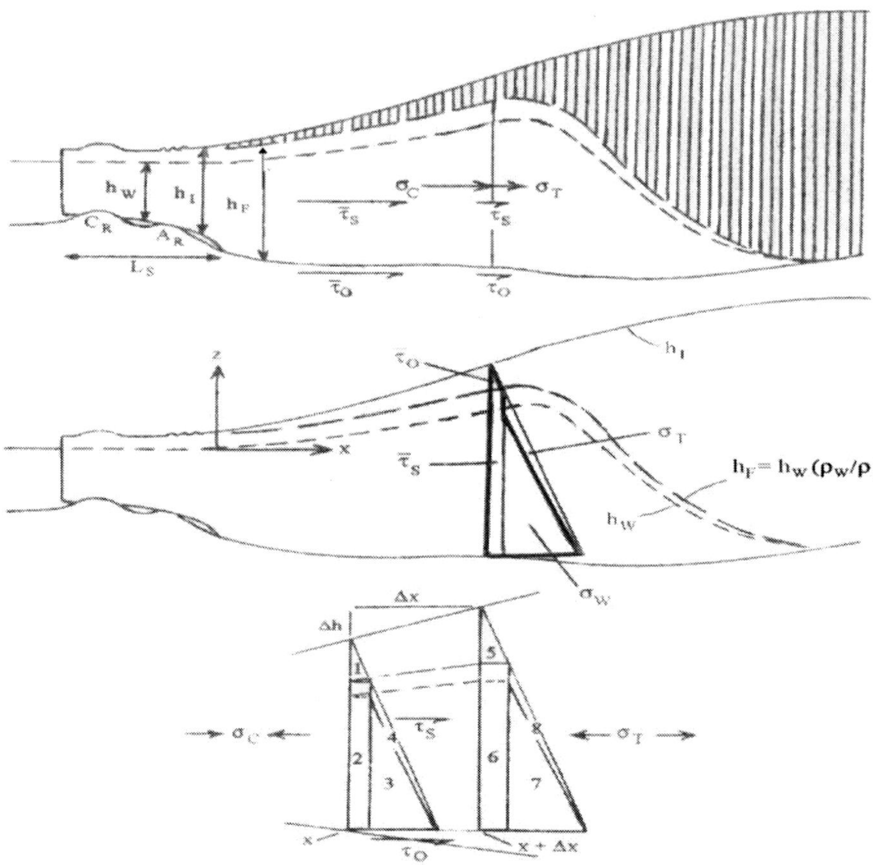

Figure 2. The geometrical force balance on an ice stream ending as a confined ice shelf. Top: Stresses that resist gravitational flow along *x*. The bed supports ice in the shaded area. Ice in the unshaded area is supported by "effective" basal water pressure $P_W$*. Middle: Gravitational forces at *x* represented as triangles and a rectangle are linked to specific resisting stresses. The area inside the thick border is linked to compressive stress $\sigma_C$. Heights $h_I$, $h_W$, and $h_F$ are measured from the bed for $x > 0$. Bottom: Resisting stresses and gravitational forces along $\Delta x$. Resisting and gravitational forces are balanced along $x$ and $\Delta x$, see Hughes (2009a). From Hughes (2009a).

Within that triangle are areas linked to resisting forces, with the area inside the bold border linked to compressive force $\sigma_C h_I$ and the remaining small triangular area linked to tensile force $\sigma_T h_I$.

This force balance gives:

$$\tfrac{1}{2}\, P_I = \overline{P}_I = \sigma_C + \sigma_T \tag{2}$$

Note that $\sigma_C >> \sigma_T$ because area 1+2+3 enclosed by the bold border greatly exceeds triangle area 4, so a minor downslope change in resistance to ice flow causes a small change in $\sigma_C$ but a large change in $\sigma_T$ because $P_I$ is initially unchanged. This illustrates the ability of $\sigma_T$ to pull more ice out of ice sheets for a small decrease in downslope resistance to ice flow (Hughes, 1992).

The bottom panel equates areas 1, 2, and 3 with compressive force $\sigma_C h_I$, triangular area 4 to tensile force $\sigma_T h_I$, triangular area 3 to water-buttressing force $\sigma_W h_I$, area 3+4 to flotation force $\sigma_F h_I$, the difference between triangular areas 5 and 4 to basal drag force $\tau_O \Delta x$, and the difference between rectangular areas 6 and 2 to side drag force $2\tau_S \Delta x$ for two sides. Balancing these longitudinal forces as $\Delta x \rightarrow 0$ gives (Hughes, 2011):

$$\sigma_T = \tfrac{1}{2}\rho_I g h_I \left(1 - \rho_I / \rho_W\right)\phi^2 = \overline{P}_I \left(1 - \rho_I / \rho_W\right)\phi^2 \tag{3}$$

$$\sigma_C = \tfrac{1}{2}\rho_I g h_I \left[1 - \left(1 - \rho_I / \rho_W\right)\phi^2\right] = \overline{P}_I \left[1 - \left(1 - \rho_I / \rho_W\right)\phi^2\right] \tag{4}$$

$$\sigma_W = \tfrac{1}{2}\rho_I g h_I \left(\rho_I / \rho_W\right)\phi^2 = \overline{P}_I \left(\rho_I / \rho_W\right)\phi^2 \tag{5}$$

$$\sigma_F = \sigma_T + \sigma_W = \overline{P}_I \phi^2 \tag{6}$$

$$\tau_O = \tfrac{1}{2}\rho_I g h_I \left(1 - \phi\right)\left[\alpha_I - \partial\left(h_I \phi\right)/\partial x\right] \approx P_I \left(1 - \phi\right)^2 \alpha \tag{7}$$

$$\tau_S = 2\rho_I g h_I \left(w_I / h_I\right)\left[\phi \alpha_I - \left(1 - 2\phi\right)\partial\left(h_I \phi\right)/\partial x\right] \approx P_I \left(w_I / h_I\right)\phi\left(1 - \phi\right)\alpha \tag{8}$$

$$\partial(\sigma_F h_I)/\partial x = \partial(\sigma_T h_I)/\partial x + \partial(\sigma_W h_I)/\partial x = P_I(\rho_I/\rho_W)\phi\alpha_W \quad (9)$$

Here $\Delta h/\Delta x \rightarrow \alpha$ is the ice surface slope, $\Delta h_I/\Delta x \rightarrow \alpha_I$ is the ice thickness gradient, and $\Delta h_W/\Delta x \rightarrow \alpha_W$ is the gradient of basal water height giving effective basal water pressure $P_W{}^*$ resisting gravitational ice flow as $\Delta x \rightarrow 0$. Water buttressing stress $\sigma_W$ in ice is due to $P_W{}^*$ in the force balance $\sigma_W h_I = P_W{}^* h_W$. Flotation stress $\sigma_F$ in ice is due to $\sigma_W + \sigma_T$ in the force balance $\sigma_F h_I = \sigma_W h_I + \sigma_T h_I$. These are real stresses. They are inaccessible using continuum mechanics in conventional ice-sheet models, but they visibly emerge from the geometrical force balance in the holistic ice-sheet model.

The longitudinal force balance equates gravitational driving force gradient $\Delta(\overline{P}_I h_I)/\Delta x = P_I\alpha$ as $\Delta x \rightarrow 0$ from the difference between area 5+6+7+8 and area 1+2+3+4 in incremental length $\Delta x$ with the areas inside these triangles linked to resisting stresses $\tau_O$ and $\tau_S$ and flotation force gradient $\partial(\sigma_F h_I)/\partial x$ to obtain (Hughes, 2011):

$$P_I\alpha = \tau_O + 2\tau_S(h_I/h_W) + \partial(\sigma_F h_I)/\partial x \quad (10)$$

Equation (10) is satisfied using substitutions from Equations (7), (8) and (9).

Now approximate bed topography with an up-down staircase in which $\Delta x$ is the constant step length and $\pm\Delta h_B$ is the variable gain or loss in step height. A normal stress $\sigma_N$ in the direction of ice flow pushes against $-\Delta h_B$ and pulls away from $+\Delta h_B$ with force $F_N = \pm\sigma_N\Delta h_B$ compared to gravitational driving force $F_G = \overline{P}_I h_I$, so that $\sigma_N\Delta h_B/\Delta x$ and $P_I\Delta h/\Delta x$ are force gradients with $\sigma_N$ close to plastic yield stress $\sigma_O$ = 100 kPa for ice (Hughes, 1998, Figure 8.3). Then $F_N$ is much less than $F_G$ until the bed slope exceeds ±30 degrees (Hughes, 2012, Appendix E), so $F_N$ can be ignored for lesser bed slopes. Then $\Delta h = \Delta h_I$ can be used for each $\Delta x$ step.

Substituting Equations (7), (8), and (9) into Equation (10), ignoring the terms containing $\partial\phi/\partial x$, dividing by $P_I$, solving for surface slope $\alpha$, and returning to the incremental form:

$$\frac{\Delta h}{\Delta x}=\frac{\Delta(\sigma_F h_I)\Delta x}{P_I}+\frac{\tau_O}{P_I}+\frac{2\tau_S(h_I/h_W)}{P_I}$$
$$=\phi^2\left(\frac{\Delta h_I}{\Delta x}\right)_F+(1-\phi)^2\left(\frac{\Delta h_I}{\Delta x}\right)_G+2\phi(1-\phi)\frac{\Delta h}{\Delta x} \quad (11)$$

Here $\Delta h_I = \Delta h$ so $(\Delta h/\Delta x)_F$ is for the floating fraction of the ice column linked to $\sigma_F$ and $(\Delta h/\Delta x)_G$ is for the grounded fraction of the ice column linked to $\tau_O$.

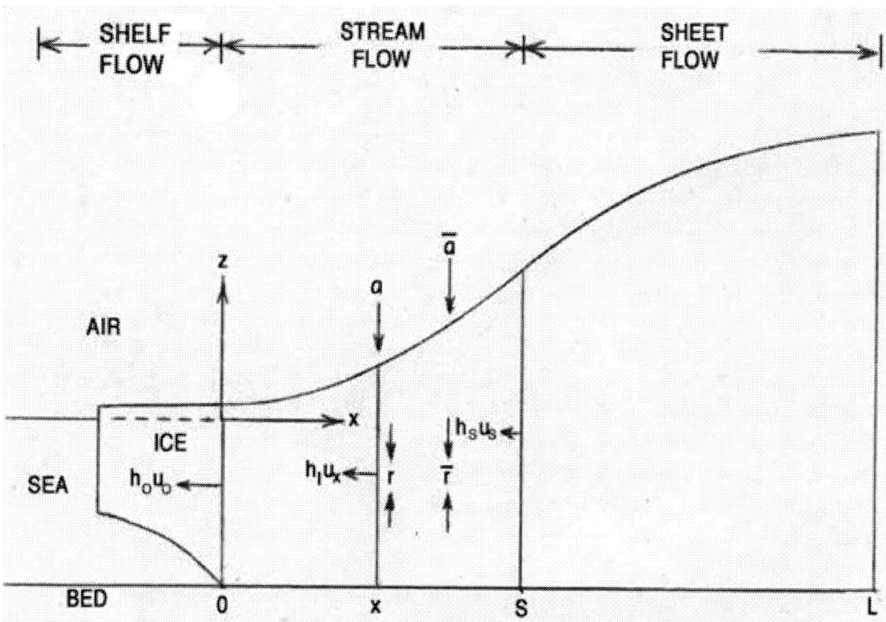


Figure 3. A longitudinal profile of an ice-sheet flowband of constant width showing components of the mass balance for sheet, stream, and shelf flow from right to left. Ice thickness $h_I$ and mean ice velocity $u_x$ are shown at $x$ positions 0, $x$, $S$, and $L$ for mean accumulation rate and ice thinning rate averaged along $x$, and rates $a$ and $r$ at $x$.

A simple mass balance is shown in Figure 3 for constant ice accumulation rate $a$ and ice thinning rate $r$, at $h_I = h_L$ where ice velocity u = 0 at the ice divide (x = L), $h_I = h_S$ where $u = u_S$ and stream flow begins ($x = S$), and $h_I = h_O$ where $u = u_O$ at the ice-shelf grounding line ($x = 0$) so that:

$$(a-r)(L-x) = h_I\,u \tag{12}$$

Using a conventional sliding law for ice (Weertman, 1957b), where B includes bed roughness and physical properties of temperate ice and m = 2 for grounded ice:

$$u = u_S = \left(\frac{\tau_O}{B}\right)^m \tag{13}$$

Here is how $(\Delta h/\Delta x)_G$ is obtained. Equate ice elevation $h$ with ice thickness $h_I$ for a horizontal bed at sea level. Combine equations (12) and (13), with $\tau_O = \rho_I\, g\, h_I\, dh_I/dx$:

$$(a-r)(L-x) = h_I u = h_I\left(\tau_O / B\right)^m = h_I\left[\left(\rho_I g h_I / B\right) dh_I / dx\right]^m \tag{14}$$

Integrating from the ice divide at $x = L$ to the ice-shelf grounding line at $x = 0$ where $h_I = h_O$, with $a$ and $r$ constant along $x$:

$$\int_O^x \left[(a-r)(L-x)\right]^{\frac{1}{m}} dx = \int_{h_O}^{h_I} \left(\rho_I g / B\right) h_I^{\frac{m+1}{m}}\, dh_I \tag{15}$$

Solving for $h_I$:

$$h_I = \left\{ h_O^{\frac{2m+1}{m}} + \frac{B}{\rho_I g}\left(\frac{2m+1}{m+1}\right)(a-r)^{\frac{1}{m}}\left[L^{\frac{m+1}{m}} - (L-x)^{\frac{m+1}{m}}\right]\right\}^{\frac{m}{2m+1}} \tag{16}$$

In Equation (16), only $a$ and $r$ determine change in $h_I$ along $x$. Now let $h_I$ vary with bed topography, using $h_I$ in Equations (14) and (16). Solve for surface slope $\alpha = dh/dx$:

$$\alpha = \frac{dh}{dx} = \frac{B}{\rho_I g h_I}\left[\frac{(a-r)(L-x)}{h_I}\right]^{\frac{1}{m}} \tag{17}$$

Taking $\tau_O = \rho_I\, g\, h_I\, \alpha$ and solving for $\tau_O$ in Equations (16) and (17):

$$\tau_O = \frac{B\left[(a-r)(L-x)\right]^{\frac{1}{m}}}{\left\{h_O^{\frac{2m+1}{m}} + \frac{B}{\rho_I g}\left(\frac{2m+1}{m+1}\right)(a-r)^{\frac{1}{m}}\left[L^{\frac{m+1}{m}} - (L-x)^{\frac{m+1}{m}}\right]\right\}^{\frac{1}{2m+1}}} \tag{18}$$

Setting $\alpha = (\Delta h/\Delta x)_G$ for ice grounded in incremental length $\Delta x$, Equations (16) and (17) give:

$$\left(\frac{\Delta h_I}{\Delta x}\right)_G = \frac{\tau_O}{\rho_I g h_I} = \frac{(B/\rho_I g)\left[(a-r)(L-x)\right]^{\frac{1}{m}}}{h_I^{\frac{m+1}{m}}}$$

$$= \frac{(B/\rho_I g)\left[(a-r)(L-x)\right]^{\frac{1}{m}}}{\left\{h_O^{\frac{2m+1}{m+1}} + \left(\frac{2m+1}{m-1}\right)\frac{B(a-r)^{\frac{1}{m}}}{\rho_I g}\left[L^{\frac{m+1}{m}} - (L-x)^{\frac{m+1}{m}}\right]\right\}^{\frac{m+1}{2m+1}}} \tag{19}$$

Using the flow law of ice (Glen, 1955), where $A$ is an ice-hardness parameter dependant on temperature and $n = 3$ for ice, $= \Delta u/\Delta x$ is the extending strain rate for stress $\sigma_T$ in floating ice, and $R$ is a dimensionless scalar that takes account of other strain rates in addition to $\dot{\varepsilon}$:

$$\dot{\varepsilon} = \dot{\varepsilon}_{xx} = \frac{\Delta u}{\Delta x} = R\left(\frac{\sigma'_{xx}}{A}\right)^n = R\left(\frac{\sigma_T}{2A}\right)^n \tag{20}$$

The mass balance in Figure 3 also applies for stream flow, with ice thickness $h_I = h_S$ where ice velocity $u = u_S$ at the beginning of stream flow ($x = S$) so that $(\Delta h/\Delta x)_F$ is:

$$\left(\frac{\Delta h}{\Delta x}\right)_F = \frac{(a-r)-h_I\dot{\varepsilon}}{u_x} = \frac{h_I(a-r)-h_I^2\dot{\varepsilon}}{h_O u_O+(a-r)\ x} \tag{21}$$

Collecting terms in Equation (11):

$$\left(1-2\phi+2\phi^2\right)\frac{\Delta h}{\Delta x} = \phi^2\left(\frac{\Delta h}{\Delta x}\right)_F + (1-\phi)^2\left(\frac{\Delta h}{\Delta x}\right)_G \tag{22}$$

Writing as a quadratic equation:

$$\left[2\left(\frac{\Delta h}{\Delta x}\right)-\left(\frac{\Delta h}{\Delta x}\right)_F-\left(\frac{\Delta h}{\Delta x}\right)_G\right]\phi^2-\left[2\left(\frac{\Delta h}{\Delta x}\right)-2\left(\frac{\Delta h}{\Delta x}\right)_G\right]\phi+\left[\left(\frac{\Delta h}{\Delta x}\right)-\left(\frac{\Delta h}{\Delta x}\right)_G\right]=0 \tag{23}$$

Setting $C_1 = (\Delta h/\Delta x)$, $C_2 = (\Delta h/\Delta x)_F$, and $C_3 = (\Delta h/\Delta x)_G$ and solving for $\phi$:

$$\phi = \frac{(2C_1-2C_3)\pm\left[(2C_1-2C_3)^2-4(C_1-C_3)(2C_1-C_2-C_3)\right]^{1/2}}{2(2C_1-C_2-C_3)} \tag{24}$$

In a flowline solution, $w_I = 0$ so $\tau_S = 0$. Yet side drag remains and contributes to the ice elevation needed to overcome resistance to ice flow, so it must be taken into account in some way. Our way is to enlarge $\tau_O$ to effective basal shear stress $\tau_O{*}$ linked to areas 5+6 minus areas 1+2 as incremental length $\Delta x \rightarrow 0$ in Figure 2. Then $\tau_O{*}$ is:

$$\tau_O{*} = \rho_I g h_I\left(1-\phi^2\right)\Delta h/\Delta x - \rho_I g h_I^2 \phi\Delta\phi/\Delta x \tag{25}$$

and the longitudinal force balance becomes, ignoring the $\Delta\phi/\Delta x$ term (Hughes, 2011):

$$\frac{\Delta h}{\Delta x} = \frac{\Delta(\sigma_F h_I)/\Delta x}{P_I} + \frac{\tau_O{*}}{P_I} = \phi^2\left(\frac{\Delta h}{\Delta x}\right)_F + \left(1-\phi^2\right)\left(\frac{\Delta h}{\Delta x}\right)_G \tag{26}$$

Collecting terms:

$$\left[\left(\frac{\Delta h}{\Delta x}\right)_G - \left(\frac{\Delta h}{\Delta x}\right)_F\right]\phi^2 - \left[\left(\frac{\Delta h}{\Delta x}\right)_G - \left(\frac{\Delta h}{\Delta x}\right)\right] = 0 \qquad (27)$$

Solving for $\phi$:

$$\phi = \pm\left[\frac{C_3 - C_1}{C_3 - C_2}\right]^{1/2} \qquad (28)$$

In Equations (24) and (28), the correct solution puts $\phi$ in the range $0 \leq \phi \leq$ 1.

## Application to Jakobshavn Isbrae

Jakobshavn Isbrae becomes afloat in Jakobshavn Isfjord, which empties into Disko Bugt on the west coast of Greenland at 69° 10' N. It has been observed longer than any other ice stream on the planet, and has traditionally been the fastest, moving about 7 km/a at its calving front (Weidick et al., 1990; Weidick, 2007). The calving front and grounding line of its floating ice tongue began a retreat of 30 km at the end of the Little Ice Age in Greenland in 1850, when the Industrial Revolution began and contributed to global warming (Sohn et al.,1998).

The first glaciological study was a photogrammetric determination of ice elevations and velocities over an area of 10,000 km$^2$ of ice converging on Jakobshavn Isflord (Brecher, 1986, Fastook et al., 1995). They found that grounded ice converges strongly as it enters Jakobshavn Isflord, so the flowband width decreases sharply. If the flowband width is taken to be constant in an ice-stream model, the flux of converging ice must be distributed over the flowband surface as an ice accumulation rate. Actual accumulation and ablation rates over this region were made by Echelmeyer et al. (1992).

Photogrammeric measurements were subsequently made in much greater detail over the floating ice tongue by Prescott et al. (2003). We will call the floating tongue an ice shelf because it is supplied by two ice streams, one being the main trunk of Jakobshavn Isbrae which continues eastward for some 100 km into the Greenland Ice Sheet following what is informally called Gogineni Gorge, see Figure 4, and the other being a much smaller local ice stream which enters Jakobshavn Isfjord from the northeast. By the time ice reaches the calving front, the smaller ice stream supplies only about 10 percent of the calving front and the main trunk retains almost a constant width from its grounding line to the calving front (Prescott et al., 2003). Ice velocities do not change much over this distance, perhaps due to resistance from side shear along the fjord walls, to local grounding near the calving front that produces ice rumples on the surface (length $L_S$ and area $A_R$ in Figure 1), substantial ice spilling over the south fjord wall and terminating as a grounded ice lobe, and high rates of surface and especially basal melting. The ice shelf rapidly disintegrated in 1997, probably for a variety of reasons. It was so heavily crevassed its dynamics may have been understood better using the theory of granular materials passing through a funnel, rather than using continuum mechanics (Johnson et al., 2004). Grains form "bridges" that arc upflow and retard ice flow, but when bridges collapse under the pressure of moving ice, they lose all cohesion, disperse catastrophically, and allow a sudden increase in discharge velocity.

Zwally et al. (2002) found that surface meltwater reached the bed by way of crevasses and moulins near the head of the smaller local ice stream, and this coincided with an increase in ice velocity during the most intense hours of sunlight. This probably occurs generally in the lower ablation zone of the Greenland Ice Sheet, and has been called The Zwally Effect. Iken et al. (1993) drilled three holes into Gogineni Gorge in the main trunk of Jakobshavn Isbrae and found a layer of temperate ice about 300 m thick above the bed. This ice can still be frozen to the bed if latent heat is not supplied to melt the ice. For example, if all ice crystals are taken to be cubes of the same size, and dislocations due to ice deformation pile up at boundaries of these crystal grains, supplying strain energy that can be converted into latent heat, melting begins first at points where eight grains meet, then progresses along lines where four grains meet, then spreads over areas where two grains meet, and finally melts the grains. This progression of melting from points to lines to areas to volumes allows a wide range of water to be stored in temperate ice. Temperate ice is soft, especially if it contains substantial meltwater, and this alone could account for the fast velocity of Jakobshavn Isbrae. In addition, if gravity supplies this internal meltwater to the bed, the water can drown bedrock bumps that pin ice to a hard bed or supersaturate and mobilize a soft bed consisting of sediments or till.

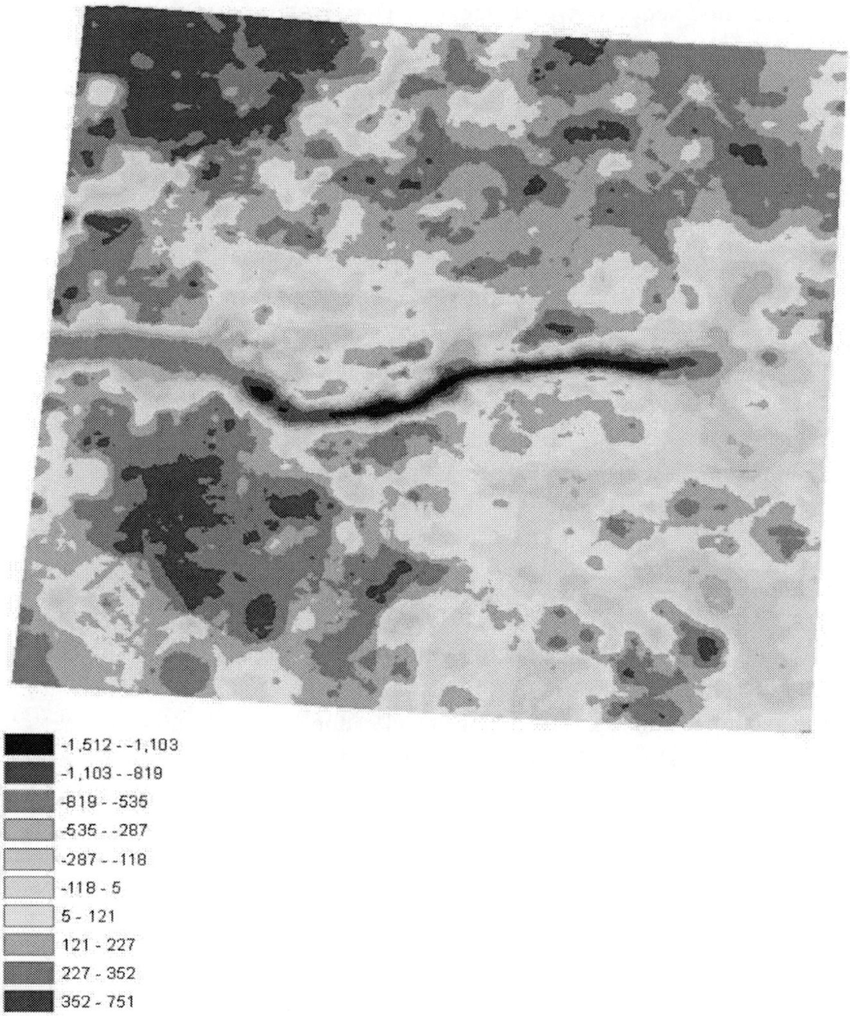


Figure 4. The bedrock map of Jakobshavn Isbræ produced from airborne radar profiles by the Center for Remote Sensing of Ice Sheets (CReSIS) at the University of Kansas. We used a digitized version for the central flightline along Gogineni Gorge for all data sets. The map is 100 km on each side.

Both processes can account for the fast velocity of Jakobshavn Isbrae. Thomas (2004) made the first attempt to model the recent doubling of velocity and resulting surface lowering of Jakobshavn Isbrae following catastrophic disintegration of its buttressing ice shelf. His approach was similar to ours, except that he considered floating fraction $\phi \approx 1$ all the way up Gogineni

Gorge. He modeled the whole ice drainage system of Jakobshavn Isbrae, including the ice catchment area that is about 7 percent of the Greenland Ice Sheet. This required taking diverging and converging flow into account, diverging from the ice divide and converging on Jakobshavn Isbrae. Because ice-bed decoupling was almost total up Gogineni Gorge, Thomas (2004) obtained a faster upflow response due to the loss of downflow buttressing when the ice shelf disintegrated. Thomas et al. (2011) conluded that this process, part of The Jakobshavn Effect (Hughes, 1986), could occur generally for all Greenland ice streams if climate continues to warm.

In our modeling attempt, we chose to model Jakobshavn Isbrae using a version of the flowline model and Equation (28) instead of the flowband model leading to Equation (24).

Our reasons are purely practical. Prasad Gogineni (personal communication, 2011) provided us with a radar flightline down the center of Gogineni Gorge, see Figure 4, which gave us continuous surface and bed topography down the centerline of Jakobshavn Isbrae. Side shear vanishes along the centerline of ice steams, so we could ignore side sear stress $\tau_S$. Also, Gogineni Gorge is very deep and it has a relatively constant width along its entire length. Laterally converging flow does not affect flow of this deep ice. Converging ice flow above the gorge can be accommodated by increasing the surface accumulation rate $a$ in Equations (19) and (21).

Equations (19) and (21) includes $(\Delta h/\Delta x)_F$ for floating fraction $\phi$ of ice in our model, linked to longitudinal strain rate as seen in Equation (21), and also the flow law of ice given by Equation (20) which links to $\sigma_T$ given by Equation (3) and $R$ given by (Hughes et al., 2011):

$$R = \left[1 + \left(\frac{\dot{\varepsilon}_{yy}}{\dot{\varepsilon}_{xx}}\right) + \left(\frac{\dot{\varepsilon}_{yy}}{\dot{\varepsilon}_{xx}}\right)^2 + \left(\frac{\dot{\varepsilon}_{xy}}{\dot{\varepsilon}_{xx}}\right)^2 + \left(\frac{\dot{\varepsilon}_{xz}}{\dot{\varepsilon}_{xx}}\right)^2\right]^{\frac{n-1}{2}} \tag{29}$$

Where $\dot{\varepsilon}_{xx}$, $\dot{\varepsilon}_{yy}$, $\dot{\varepsilon}_{xy}$, and are strain rates associated with longitudinal extension, lateral compression, side drag, and basal drag, respectively. Lateral compression occurs in converging thin upper ice but not in thick lower ice occupying Gogineni gorge. There is no lateral shear down the centerline, and there is little basal shear if the bed is wet and $\phi$ is high. So is the dominant strain rate and $R \approx 1$ for $n = 3$. The longitudinal strain rate is therefore, using Equation (3) for $\sigma_T$, following Hughes et al. (2011):

$$\dot{\varepsilon} = \dot{\varepsilon}_{xx} = (\sigma_T / 2A)^n = \left[ (\rho_I g h_I / 4A)(1 - \rho_I / \rho_W)\phi^2 \right]^n$$
$$= \left[ (\rho_I g h_I / 4A)(1 - \rho_I / \rho_W) - (\sigma_B / 2A) \right]^n \qquad (30)$$

Here $\sigma_B$ is a back-stress due to buttressing by the ice shelf in Jakobshavn Isfjord and given by:

$$\sigma_B = f_B \left[ \tfrac{1}{2} \rho_I g h_O (1 - \rho_I / \rho_W) \right] \qquad (31)$$

where $f_B$ is a buttressing fraction with $f_B = 0$ for no buttressing and $f_B = 1$ for full buttressing.

Equations (30) and (31) allow two possibilities for variations along $x$. If $\phi$ is also allowed to vary along $x$ in Equation (30) the solution is:

$$\dot{\varepsilon} = \left[ (\rho_I g h_I / 4A)(1 - \rho_I / \rho_W)\phi^2 \right]^n \qquad (32)$$

with $\phi$ at $x = 0$ being a measure of ice-shelf buttressing such that $\phi = 1$ if the ice shelf has disintegrated. If $\phi$ is replaced by ice-shelf buttressing at $x = 0$, then Equation (31) requires that:

$$\dot{\varepsilon} = \left[ \tfrac{1}{2} \rho_I g h_I (1 - \rho_I / \rho_W) \right]^n \left[ 1 - f_B (h_O / h_I) \right]^n / (2A)^n \qquad (33)$$

Equation (33) emphasizes that ice-shelf buttressing is transmitted upstream. With either Equation (32) or Equation (33) substituted for in Equation (21), we see that $(\Delta h/\Delta x)_F$ varies with either $\phi^6$ or $[1 - f_B(h_O/h_I)]^3$ for $n = 3$. Both possibilities will be considered. In the case of Equation (33), $0 \leq f_B \leq 1$ is chosen to match the observed $h_O$ at the ice-shelf grounding line. When the ice shelf has disintegrated, $f_B = 0$ is expected. Equation (33) is not so different from one used by Thomas (2004) in modeling the ongoing surge of Jakobshavn Isbrae following disintegration of its buttressing ice shelf in Jakobshavn Isfjord.

Before our approach can be applied to Jakobshavn Isbrae, it is necessary to address a concern that arises from approximating bed topography using an up-down staircase. As shown by Hughes (2012, Appendix E), this is acceptable for bed slopes not exceeding 30 degrees in the force balance. In the mass-balance, however, the staircase produces abrupt velocity changes between $\Delta x$ steps, an abrupt decrease at "down" steps and increase at "up" steps, because ice thickness $h_I$ increases and decreases abruptly between steps in the mass balance.

Therefore gradients $(\Delta h/\Delta x)_G$ and $(\Delta h/\Delta x)_F$ used as surface slopes in Equations (19) and (21) for the force balance should be retained as ice-thickness gradients $(\Delta h_I/\Delta x)_G$ and $(\Delta h_I/\Delta x)_F$ in the mass balance. This is done by setting:

$$\frac{\Delta h}{\Delta x} = \frac{\Delta h_I}{\Delta x} - \frac{\Delta h_B}{\Delta x} \tag{34}$$

where $\Delta h_B/\Delta x$ is the bed slope. Bed slopes are known along Jakobshavn Isbrae so this information should be used.

Since $\Delta h_B/\Delta x$ due to changing bed topography represented by $h_B$ did not affect the derivation of Equation (19), see Denton and Hughes (1981, Chapter 5), Hughes (1998, Chapter 5), and Hughes (2012, Chapter 17), Equation (19) remains:

$$\left(\frac{\Delta h}{\Delta x}\right)_G = \frac{\left(B/\rho_I g\right)\left[(a-r)(L-x)\right]^{\frac{1}{m}}}{\left\{h_O^{\frac{2m+1}{m+1}} + \left(\frac{2m+1}{m+1}\right)\frac{B(a-r)^{\frac{1}{m}}}{\rho_I g}\left[L^{\frac{m+1}{m}} - (L-x)^{\frac{m+1}{m}}\right]\right\}^{\frac{m+1}{2m+1}}} = K_1 \tag{35}$$

where $K_1 = C_3$ in Equation (23).

Derivation of Equation (35) includes $\Delta h_B/\Delta x$. For the mass balance with $h_O\, u_O$ at the ice-shelf grounding line, $x = 0$, and $h_I\, u$ at distances upstream from the grounding line in Figure 3, ice thinning rate $r$ in time $t$ is:

$$\frac{\delta h_I}{\delta t} = r = a - \frac{\Delta(h_I u)}{\Delta x} = a - h_I\frac{\Delta u}{\Delta x} - u\frac{\Delta h_I}{\Delta x} \tag{36}$$

Solving for ice thickness gradient $\Delta h_I/\Delta x$; using Equation (36):

$$\left(\frac{\Delta h_I}{\Delta x}\right)_I = \frac{a-r}{u} - \frac{h_I(\Delta u/\Delta x)}{u} = \frac{h_I(a-r) - h_I^2\dot{\varepsilon}}{h_O u_O + (a-r)x} = \left(\frac{\Delta h}{\Delta x} + \frac{\Delta h_B}{\Delta x}\right)_I \tag{37}$$

where ice flux $h_I u = h_O u_O + (a - r)x$ and $= \Delta u/\Delta x$ is the longitudinal strain rate, both at $x$ in Figure 3. The ice surface slope for floating ice is then:

$$\left(\frac{\Delta h}{\Delta x}\right)_F = \left(\frac{\Delta h_I}{\Delta x}\right)_F - \frac{\Delta h_B}{\Delta x} = \frac{h_I(a-r) - h_I^2\dot{\varepsilon}}{h_O u_O + (a-r)x} - \frac{\Delta h_B}{\Delta x} \tag{38}$$

For an ice shelf, $\Delta h_B/\Delta x$ is the bottom surface of floating ice and must therefore satisfy the buoyancy condition:

$$\left(\frac{\Delta h_B}{\Delta x}\right)_F = \left(\frac{\Delta h_I}{\Delta x} - \frac{\Delta h}{\Delta x}\right) = \frac{\Delta}{\Delta x}\left(\frac{h}{1-\rho_I/\rho_W}\right) - \frac{\Delta h}{\Delta x} = \left[\frac{\rho_I/\rho_W}{1-\rho_I/\rho_W}\right]\frac{\Delta h}{\Delta x} \tag{39}$$

where positive $\Delta h_I/\Delta x$ and $\Delta h/\Delta x$ require negative $\Delta h_B/\Delta x$. Note that in Equation (38) has two solutions, Equation (32) and Equation (33).

We have applied three models, based on the geometrical force balance, to Jakobshavn Isbrae: *GFB*1, *GFB*2, and *GFB*3, where *GFB*1 uses Equation (33) for $\dot{\varepsilon}$, *GFB*2 uses Equation (32) for $\dot{\varepsilon}$, and *GFB*3 shows the consequences of ignoring bed slope by setting $\Delta h_B/\Delta x = 0$. With Equation (33), similar to Thomas (2004), Equation (26) including $\Delta h_B/\Delta x$ becomes:

$$\frac{\Delta h}{\Delta x} = \left(\frac{\Delta h}{\Delta x}\right)_F \phi^2 + \left(\frac{\Delta h}{\Delta x}\right)_G \left(1-\phi^2\right) = \left[\left(\frac{\Delta h_I}{\Delta x}\right)_F - \frac{\Delta h_B}{\Delta x}\right]\phi^2 + \left(\frac{\Delta h}{\Delta x}\right)_G \left(1-\phi^2\right) \tag{40}$$

Using Equation (33) for $\dot{\varepsilon}$ gives the expression for *GFB*1:

$$\frac{\Delta h}{\Delta x} = \left[\frac{\Delta h_I}{\Delta x} - \frac{\Delta h_B}{\Delta x}\right]\phi^2 + K_1\left(1-\phi^2\right) = \left[\frac{h_I(a-r) - h_I^2\dot{\varepsilon}}{h_O u_O + (a-r)} - \frac{\Delta h_B}{\Delta x}\right]\phi^2 + K_1\left(1-\phi^2\right)$$

$$= \left[\frac{h_I(a-r) - h_I^2\left[\tfrac{1}{2}\rho_I g h_I\left(1-\rho_I/\rho_W\right)\right]^n\left[1-f_B\left(h_O/h_I\right)\right]^n/(2A)^n}{h_O u_O + (a-r)x} - \frac{\Delta h_B}{\Delta x}\right]\phi^2 + K_1\left(1-\phi^2\right)$$

$$= \left(K_2 - \Delta h_B/\Delta x\right)\phi^2 + K_1\left(1-\phi^2\right) \tag{41}$$

Uisng Equation (32) for gives the expression for *GFB*2:

$$\frac{\Delta h}{\Delta x} = \left[\frac{\Delta h_I}{\Delta x} - \frac{\Delta h_B}{\Delta x}\right]\phi^2 + K_1\left(1-\phi^2\right) = \left[\frac{h_I(a-r) - h_I^2\dot{\varepsilon}}{h_O u_O + (a-r)x} - \frac{\Delta h_B}{\Delta x}\right]\phi^2 + K_1\left(1-\phi^2\right)$$

$$= \left[\frac{h_I(a-r)}{h_O u_O + (a-r)x}\right]\phi^2 + \left[\frac{-h_I^2\left[\left(\rho_I g h_I/4A\right)\left(1-\rho_I/\rho_W\right)\phi^2\right]^n}{h_O u_O + (a-r)x}\right]\phi^2 - \left(\frac{\Delta h_B}{\Delta x}\right)\phi^2 + K_1\left(1-\phi^2\right)$$

$$= \left[\frac{h_I(a-r)}{h_O u_O + (a-r)x} - \frac{\Delta h_B}{\Delta x}\right]\phi^2 + \left[\frac{-h_I^{n+2}\left[\left(\rho_I g/4A\right)\left(1-\rho_I/\rho_W\right)\right]^n}{h_O u_O + (a-r)x}\right]\phi^{2n+2} + K_1\left(1-\phi^2\right)$$

$$= \left(K_3 - \Delta h_B/\Delta x\right)\phi^2 + K_4\phi^{2n+2} + K_1\left(1-\phi^2\right) \tag{42}$$

The consequences of ignoring bed slope are found by setting $\Delta h_B/\Delta x = 0$ in Equations (41) and (42). This is *GFB*3.

The recent and sustained surge of Jakovshavn Isbrae began when its buttressing ice shelf in Jakobshavn Isfjord disintegrated suddenly in 1997. This has been attributed to warm ocean water entering the fjord and passing under the ice shelf (Holland et al., 2008), when the melting rate was already 109 m/a, the highest measured for any ice shelf on Earth (Prescott et al., 2003). There was also substantial surface melting, with most surface meltwater reaching the bed through crevasses (Hughes 1986; Echelmeyer et al., 1992).

Therefore it is prudent to include the ice shelf in our model. This was done by moving the origin of x, y, z coordinates to the calving front of the ice shelf and using Equation (39) for evaluating the buoyancy condition using known top and bottom slopes measured on the ice shelf.

## RESULTS FOR JAKOBSHAVN ISBRAE

For our analysis of Jakobshavn Isbræ, we used ice-surface data sets from 1985, 1993, 2003 and 2006. This gave us four sequential time shots and three sequential time steps with $\phi$ distributions along the central flowline. The data sets represent approximately the last 70 km along the central flowline of the Jakobshavn Isbræ ice drainage system, which has a total length of about 570 km from the ice divide to the calving front (Bindschadler, 1984). Surface data sets have been collected by different means, so they have different resolutions. Surface data for 1985 were obtained using gridded data from 1985 and 1986 measured photogrammerically, with an accuracy of about 2 m (Brecher, 1986). Surface data for 1993 and 2003 were collected by NASA glaciologists using an airborne laser scanning system with an accuracy of 0.1 m. Surface data for 2006 along the center flowline were collected by the Center for Remote Sensing of Ice Sheets (CReSIS) at the University of Kansas using airborne radar with an accuracy of about 1 m. CReSIS also provided bed topography obtained by radar sounding along the center flowline and shown in Figure 4. This topography is used in all data sets and is accurate within a few tens of meters.

Until approximately 1997 Jakobshavn Isbræ still ended as an ice shelf approximately 7 km long. The length of the ice shelf in 1985 and 1993 was determined from surface data using a flotation criterion in which the depth of basal ice below sea level is $b = -h/(1 - \rho_W / \rho_I)$ for surface elevation $h$. The position of the grounding line was determined when bedrock depth $h_B$ was reached by basal ice.

Ice deformation is largely determined by ice hardness parameter $A$ in Equation (20) that varies with ice temperature. There is quite some variation of ice temperatures in Jakobshavn Isbræ depending on time and place. Borehole studies showed a temperature variation with depth from -2 °C to a minimum of -22 °C (Iken et al., 1993; Lüthi et al., 2002).

Thomas et al. (2011) preferred ice temperatures from -4 °C to -6 °C for Jakobshavn. We fitted $A$ to calculated strain rates for the ice stream determined in 1985 (Fastook et al., 1995), giving $A = 1.4\times10^8$ Pa $s^{1/3}$ corresponding to a temperature of -15 °C (Paterson, 1994) and kept $A$ constant through the different time shots.

As ice motion is very fast and the bed is likely to be thawed, we assumed ice is sliding. Whether this is due to high deformation in a temperate ice layer (Lüthi et al., 2009), high creep deformation in a layer with ice crystals having vertically orientated c-axes (Horgan et al., 2008), or ice sliding over bedrock

(Weertman, 1957b) is not important. Hence in Equation (13) we use a value of $B$ that simulates fast-moving ice over bedrock, $B = 1.123 \times 10^7$ Pa m$^{1/2}$ s$^{-1/2}$ (Reusch and Hughes, 2003).

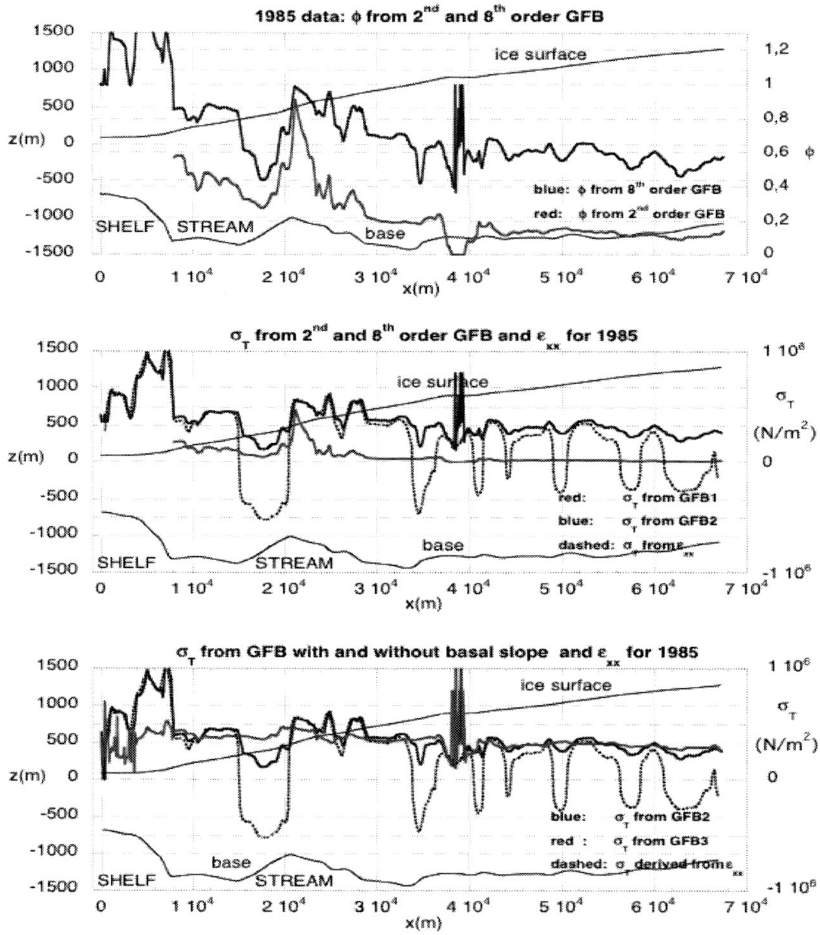


Figure 5. Results for the three different versions of the GFB using data from 1985 for Jakobshavn Isbræ. Profiles of basal and surface ice, shown in black, are quantified on the left vertical axis, while $\phi$ or $\sigma_T$ is quantified on the right vertical axis. The top graph investigates the effect of including $\phi$ in $\sigma_T$ by comparing GFB1 in red, when $\sigma_T$ is independent of $\phi$, with GFB2 in blue, when $\sigma_T$ depends on $\phi$. The middle graph compares $\sigma_T$ from GFB1 in red with $\sigma_T$ from GFB2 in blue. The black dashed graph is $\sigma_T$ calculated from $\dot{\varepsilon}_{xx}$ and is taken as the correct $\sigma_T$. The bottom graph investigates the effect of including the basal slope in the floating part of the GFB by comparing GFB2 with GFB3, which is GFB2 when $\Delta h_B/\Delta x = 0$.

Outlet ice velocity $u_O$ at $h_O$ varied from one data set to another. From Joughin et al. (2004) we know the outlet velocities from 1985, 1993 and 2003 varied from 6700 m $a^{-1}$, 5600 m $a^{-1}$, to 12600 m $a^{-1}$, values we used for those data sets. For 2006 we assumed an outlet velocity of 14000 m $a^{-1}$ based on an annual linear increase of 5% (Joughin et al., 2008).

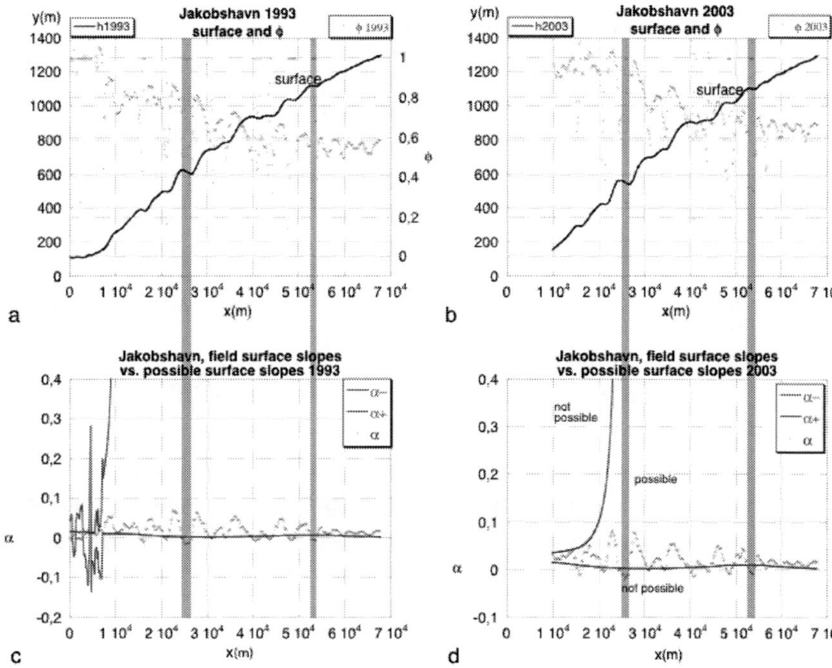


Figure 6. A comparison between $\phi$ and the ice surface (top graphs) and a comparison between the surface slope limits of the model and the actual surface slopes of Jakobshavn Isbræ in 1993 and 2003 (bottom graphs). When the model can no longer reproduce the measured surface slope (mostly when the slope is negative) the calculated value of $\phi$ is meaningless. (a) Shown here are the calculated $\phi$ values (red crosses) and ice surface surface of 1993 plotted along the flow line. The surface is quantified against the left vertical axis and $\phi$ is quantified against the right vertical axis. (b) The same data representaion as in (a) but now for the 2003 data. (c) Shown here are the minimum (red continuous line) and maximum (green continuous line) surface slope limits of the model and the actual surface slopes from the field data (blue crosses) from 1993. (d) The same data representation as in (c) but now for the 2003 data. In the shaded areas the surface slope of the field data cannot be reproduced by the model. In the bottom graphs the blue crosses lie outside the red and green slope limits of the model. As a result, $\phi$ shown in these graphs is meaningless here. The spiky lines at the start of the 1993 data are caused by the irregular ice-shelf profile due to many deep crevasses. In 2003 the ice shelf is gone and the terminus retreated about 10 km.

Actual estimates of accumulation rate vary between 0.28 m $a^{-1}$ to 0.59 m $a^{-1}$ (Bindschadler, 1984). For calculalting $\phi$ variations, uncertainties in accumulation rates hardly matter. We used an accumulation rate of 0.59 m $a^{-1}$.

Until 1985, Jakobshavn Isbræ apparently was close to steady state. The calving front did not move much and the terminal velocity was rather constant (Bindschadler, 1984; Pelto et al., 1989; Echelmeyer et al., 1992). Seasonality of the outlet velocity was hardly noticeable.

To determine what force balance is best, we calculated $\phi$ and $\sigma_T$ using Equations (41), (42), and (42) with $\Delta h_B/\Delta x = 0$ for GFB1, GFB2, and GFB3, respectively, all using the 1985 data set. Results for testing GFB1, GFB2, and GFB3 are shown in Figure 5, where $\phi$ or $\sigma_T$ is plotted along the centerline of Jakobshavn Isbræ. The top graph investigates the effect of using Equation (3) for $\sigma_T$ when $\phi$ =1 or $1 \geq \phi \geq 0$ by comparing GFB1 in red, given by Equation (41), with GFB2 in blue given by Equation (42). The vertical left axis displays the values of $\phi$, the horizontal and right axes represent distance from the origin, the 1985 calving front, and ice elevation above sea level. Both graphs have similar shape but GFB2 gives higher values for $\phi$. This affects $\sigma_T$ as shown in the second graph. The blue graph represents $\sigma_T$ calculated from GFB2 and the red graph from GFB1. These values are plotted on the left vertical axis. As a reference signal, $\sigma_T$ was calculated from longitudinal strain rate $\dot{\varepsilon}_{xx}$, shown as the black dashed graph, assuming $\sigma_T$ is the true tensile stress.

By using Equation (10) with $\tau_S$ = 0, we restricted ourselves to a two-dimensional flowline solution, whereas Jakobshavn Isbræ is a complex three-dimensional system. To get meaningful results, two-dimensional strain rates were calculated from the ice thickness data. Ice thickness gradient $\partial h_I/\partial x$ and the outlet ice flux were calculated assuming a steady state. Calculating strain rates in this way gives negative values for the tensile stress whenever the ice thickness gradient diverges in the flow direction. As mentioned, the ice hardness parameter $A$ was tuned such that the values of $\sigma_T$ , the blue line, from GFB2 matched the true values of $\sigma_T$, the black dashed line. Especially in the ice shelf, the GFB2 estimate is very good. This causes unrealistic values of $\phi$ for the ice shelf in the top graph where $\phi > 1$. Obviously $\sigma_T$ needs to be higher than linear shelf flow can produce. This is somewhat surprising, because the ice shelf of Jakobshavn Isbræ consisted of nearly parallel flow in 1985 (Prescott et al, 2003). So we would expect a flowline model to be suitable. Prescott et al. (2003) calculated a high surface-and-basal melting rate of 109 m/a (mostly basal) that we did not take into account. The high melting rate produces a large ice thickness gradient which overestimates $\sigma_T$. This unrealistic high value of $\sigma_T$ requires that $\phi > 1$. The bottom graph investigates the effect of including the basal slope in the GFB by comparing Equation (42) with Equation (42) when $\Delta h_B/\Delta x = 0$. On average, $\sigma_T$ is comparable in magnitude but not in shape. Including the basal slope in the GFB makes $\sigma_T$ more similar to the true tensile stress obtained from $\dot{\varepsilon}_{xx}$. Whenever the true tensile stress becomes negative, as for diverging flow, $\sigma_T$ from GFB2 follows this trend but remains positive. This is because GFB2 links the magnitude of a force to an area in Figure 2 that cannot be negative.

A second shortcoming of the GFB is it accommodates only a limited range of surface slopes. Figure 2 shows that the idealized sheet, stream, and shelf slopes along the flowline have only positive surface slopes, whereas the field data often have negative slopes too. The model cannot account for local drastic slope changes or negative slopes because $\Delta h/\Delta x$ is too strong as a variable. At these places the predicted $\phi$ values are meaningless. This problem is particularly clear for the highly resolved surface data from 1993 and 2003. Figure 6 shows the boundaries of possible surface slopes from the model versus the measured surface slopes from 1993 and 2003. The red graph is the lower limit of the model's surface slopes and the green graph the upper limit. The red crosses represent the actual surface slopes along the flow line and they fall occasionally outside the area of possible surface slopes, as marked by the shaded rectangles. Here the calculated $\phi$ values fall outside the model's range and are meaningless. Using a running mean for surface slopes can smooth the surface enough to eliminate false local variations in $\phi$.

In general GFB2 gave best results for $\sigma_T$ so Equation (42) is the equation we solved for the consecutive surface profiles of Jakobshavn Isbræ. Floating fraction $\phi$ is presented in Figure 7 and its tensile stress $\sigma_T$ is shown in Figure 8. The top graph shows the total change, from 1985 to 2006. The three graphs under this represent the sequential time steps leading to the total time interval. In all time steps, the blue graphs represent the older data set and the red graphs represent the newer data set. These values are plotted against the right vertical axis. Ice surface and bed data are plotted in black, where the dashed section is the older data.

In Figure 7 the general trend of all graphs is that $\phi \approx 1.0$ at the calving front except for the 1985 data set where $\phi$ is a bit higher, suggesting almost completely buoyant ice to the grounding line, with gradually more grounding going upstream.

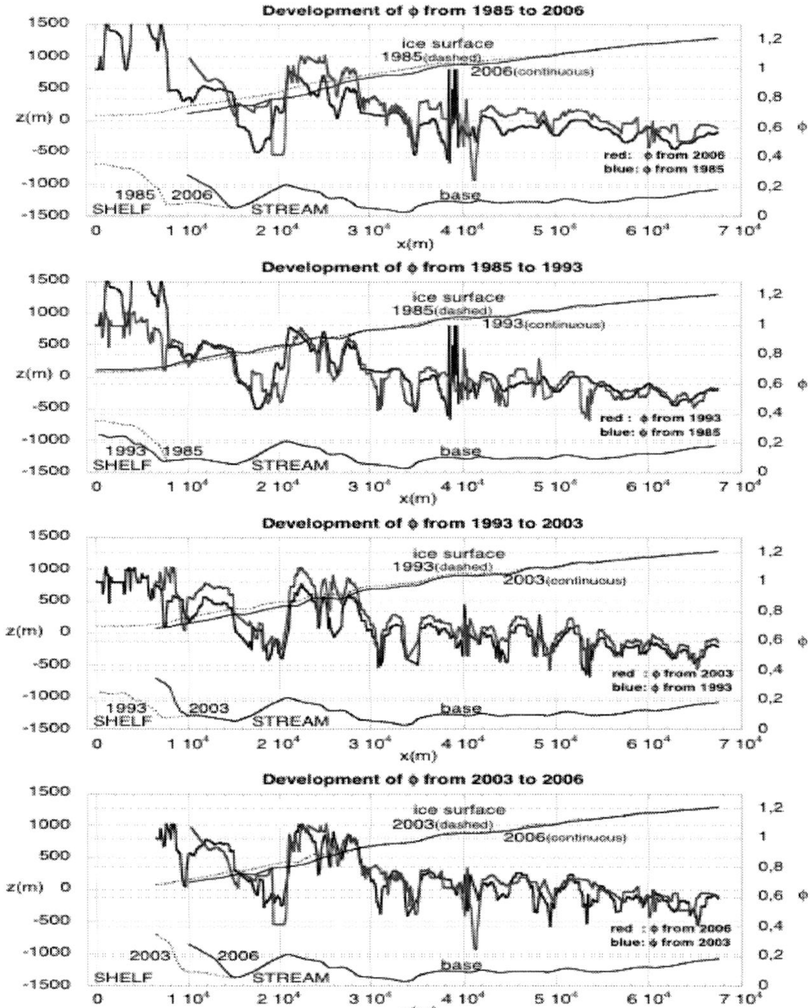


Figure 7. The development of $\phi$ using GFB2 for Jakobshavn Isbræ. The first (top) graph shows the overall development from 1985 to 2006. The blue graph shows the old values of $\phi$ and the red graph shows the new values plotted against the right vertical axis. The surface and bedrock data are plotted in black against the left vertical axis where a possible alternative bed profile is dashed. The next three graphs show the consecutive time steps, the second graph is time step 1985 to 1993, the third graph is from 1993 to 2003, and the fourth (bottom) graph is from 2003 to 2006.

Even 70 km upstream from the grounding line, $\phi \approx 0.55$. The ice stream is largely buoyant but becomes more grounded in the upstream direction. This suggests the presence of basal water throughout the modeled flowline and confirms the possibility of a temperate ice layer suggested by Lüthi et al. (2002, 2009) and Iken et al. (1993). Since Gogineni Gorge under Jacobshavn Isbræ is roughly 800 to 1500 m below sea level, a wet base can be expected. The top graph, covering the period from 1985 to 2006, shows a significant increase in $\phi$ over the entire flowline, decreasing even further upsteam, suggesting an increased buoyancy over this period. The largest increase in buoyancy took place near the ice-shelf grounding line.

If we then look at the consecutive time steps, we see that between 1985 and 1993 there is no clear change in $\phi$ except for the ice shelf. The ice shelf thickened in this period and $\phi$ decreased. Upstream from the grounding line, $\phi$ is relatively unchanged. Local differences are to be expected, as resolution of the 1993 data is much higher and thus the surface profile is different (more rugged). In general the shape and magnitude of $\phi$ is the same. By then the shelf had disintegrated.

From 1993 to 2003 we see an overall increase in $\phi$ along the entire flowline with the largest change near the calving front. The shape of the graphs is very similar but so is the resolution. The bottom graph covers the period from 2003 to 2006. The resolution of the 2006 surface data is about a meter, making a comparison difficult. There is a small increase in $\phi$ starting 13 km upstream from the grounding line, suggesting an increase in buoyancy. The higher resolution of the 1993 and 2003 surface data sets makes the $\phi$ values noisier. The surface data are bumpier and especially low surface slopes fall outside the range the model can produce.

Development of tensile stress $\sigma_T$, shown in Figure 8, is very similar to the $\phi$ development. Increasing $\sigma_T$ downstream seen in the first (top) graph is in accordance to the observed velocity increase over the period 1985 to 2006. From 1985 to 1993 not much change was observed. The first 20 km of the ice thickened somewhat, in particular for the ice shelf, reducing ice thickness gradient $\partial h_I/\partial x$. In the same period the outlet velocity reduced somewhat from 6700 m $a^{-1}$ to 5700 m $a^{-1}$ (Joughin et al., 2004). Both effects contribute to a smaller $\sigma_T$ as is registered. This slowing down has been attributed to possible jamming in the ice shelf (Johnson et al., 2004) and could explain why thickening is observed only at the front. To pinpoint this hypothesis, velocity data for 1985 and 1993 upstream from the thickening are compared in the second graph. If they remained unchanged (and the values for $\sigma_T$ suggest this), jamming may well have occurred in this period.

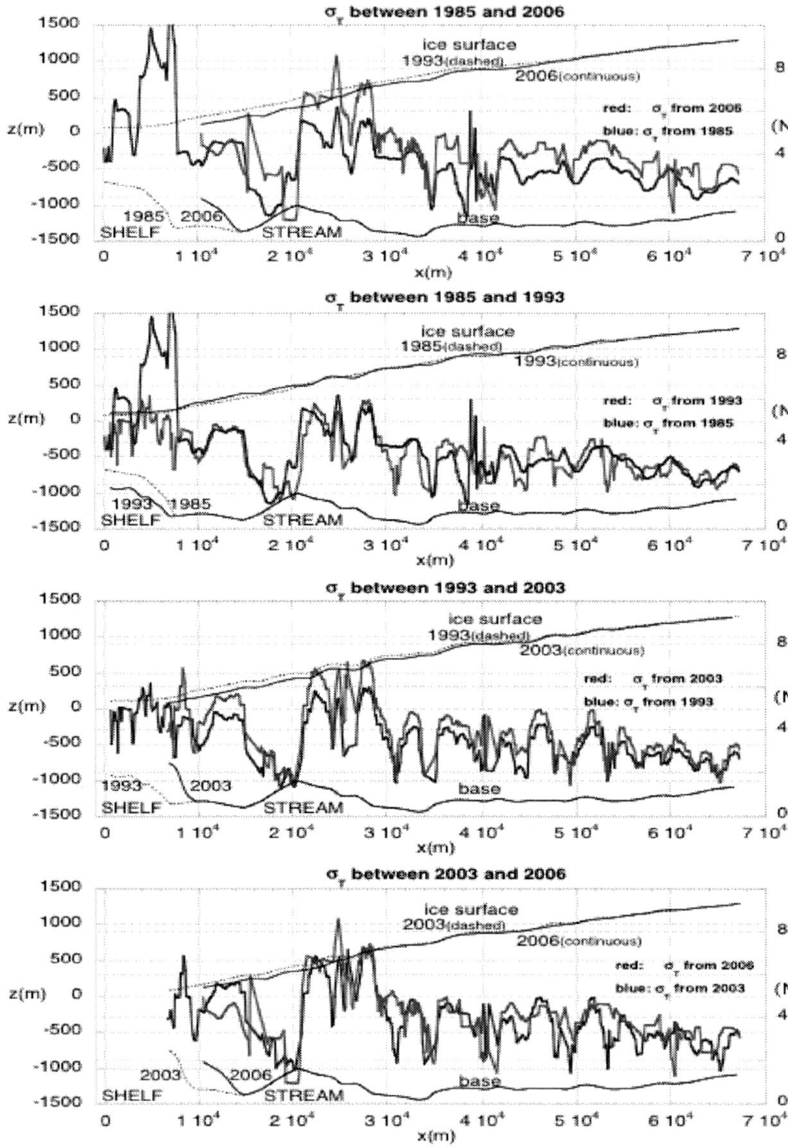


Figure 8. The development of $\sigma_T$ using GFB2 for Jakobshavn Isbrae. The first (top) graph shows the overall development from 1985 to 2006. The blue graph shows the old values of $\sigma_T$ and the red graph shows the new values plotted against the right vertical axis. The surface and bedrock data are plotted in black against the left vertical axis where a possible alternative bed profile is dashed. The next three graphs show the consecutive time steps, the second graph is time step 1985 to 1993, the third graph is from 1993 to 2002, and the fourth (bottom) graph is from 2003 to 2006.

Over the next period, from 1993 to 2003, Figure 8 shows a general increase in the value of $\sigma_T$ all along the flowline. This time interval is the period in which most changes in ice dynamics for Jakobshavn Isbræ have been documented.

After the outlet velocity slowed down somewhat in the early nineties, the ice stream started speeding up around 1998 (Joughin et al., 2008). This velocity increase came in steps, at least initially. Around 2003 the small ice shelf disintegrated (Podlech and Weidick, 2004), possibly followed by increased velocity (Thomas et al., 2003). By this time the outlet velocity increased from 5600 m $a^{-1}$ in 1992 to 12600 m $a^{-1}$ (Joughin et al., 2004) and had significant seasonal variation. The ice-shelf breakup in its turn was probably caused by an unusual influx of warmer ocean water (Holland et al., 2008). The observed increase in $\sigma_T$ took place just upslope from the grounding line and gradually decreased in the upstream direction. In general, the ice stream became more uncoupled or more shelf-like in its behavior. One reason for this is ice thinning due to an increased outlet velocity. If the ice thins, the ice overburden pressure $P_I$ decreases and thus $\phi$ and $\sigma_T$ increase. This is known as the Jakobshavn Effect (Hughes, 1986).

As shown in the fourth (bottom) graph of Figure 8, for $\sigma_T$, from 2003 to 2006 there was a slight increase in $\sigma_T$ mainly after the first 13 km. A somewhat blocky character of the 2006 data set is due to the decreased resolution and the linear extrapolation that is used between surface data points. This time step only covers three years so not as much change compared to the previous periods can be expected.

The data would let us believe that most velocity changes took place about 13 km upstream from the front, but according to Joughin et al. (2008) the velocity of Jakobshavn Isbræ has increased linearly about 5 percent per year since 2003. At the same time the seasonal variation of the outlet velocity increased by about 1000 m $a^{-1}$. This is not observed in the modeled data.

## CONCLUSION

Our holistic model was applied to transitions from sheet flow to stream flow to shelf flow along the central flowline of Jakobshavn Isbrae in Greenland. It quantifies a horizontal force balance along flow using the geometrical force balance (GFB) coupled to a simplified mass balance. The model needs little more than an outlet ice flux at the calving front of the terminal ice shelf in Jakobshavn Isfjord and profiles of both the top and bottom surfaces along the flowline. These were provided by Prasad Gogineni, Director of the Center for Remote Sensing of Ice Sheets (CReSIS) at The University of Kansas. Surface accumulation and ablation rates for sheet and stream flow, and top and bottom rates for shelf flow, have been measured but they have no major effect on the model. This justifies using a simplified mass balance equation. Jakobshavn Isbrae lies in the ablation zone of the Greenland Ice Sheet.

However ice flow is strongly convergent on Jakobshavn Isfjord in this zone, and the flowline model can take this ice flux into account by converting it into an ice precipitation flux along the flowline. This puts the flowline in an effective accumulation zone. The converging ice flux affects only the upper part of Jakobshavn Isbrae because the lower part occupies a deep extension of Jakobshavn Isfjord we have called "Gogineni Gorge" to honor Prasad Gogineni, who mapped the gorge and the whole region by radar sounding.

The GFB includes a longitudinal tensile stress that "pulls" ice out of the ice sheet and exists because of a large "floating fraction" of ice we quantify as $\phi$ and which gives stream flow its fast velocity. Three versions of the model were tested, GFB1 in which the main cause of increased flow was sudden disintegration of the buttressing ice shelf, GFB2 in which increased flow was caused mainly by increasing $\phi$ upstream due to surface meltwater reaching the bed by way of crevasses and moulins after the ice shelf disintegrated, and GFB3 which applied the model to a hypothetical condition that ignored the effect of changing bed topography upstream. The best match with observed surface velocities was provided by GFB3. However, both processes, reduced buttressing and an increasing floating fraction, are related and responsible for the doubling of discharge velocity from 1997 to 2006.

The surface of Jakobshavn Isbrae is quite bumpy, for reasons not fully understood. Some bumps have negative slopes, which will give negative values of $\phi$, and some bumps have slopes too steep to keep $\phi$ in the acceptable range $0 \leq \phi \leq 1$. Therefore a running mean of surface elevation data should be employed to keep $\phi$ in this range.

The main shortcomings of the model are (1) it applies to a limited range of surface and bed slopes, (2) flowlines cannot treat diverging and converging ice flow, (3) the effects of side drag in ice steams has to be absorbed into basal drag along flowlines, (4) the model has no provision for changing flowlines through time, (5) the energy balance is not included so temperatures in the ice and thermal conditions at the bed are not obtained, (6) the model is not readily adaptable to the map plane, and (7) the flowband version of the model would have to convert the force triangles and rectangle in Figure 1 into volumes for flowbands of variable width. Our results show the surge of Jakobshavn Isbrae that began in 1997, when its buttressing ice shelf disintegrated, has not abated and in fact is accelerating. This is confirmed by ongoing field observations (Thomas et al., 2003, 2011). The main issue we wanted to address was the short-term stability of the Greenland Ice Sheet, as measured over the next 300 years (Hughes, 2004; Pfeffer, 2008). From our modeling results, there is no inherent reason why the ongoing surge of Jakobshavn Isbrae cannot continue well into the future and be joined by surges of other important Greenland ice streams, as predicted by positive feedback mechanisms that make The Jakobshavn Effect propagate northward along both the east and west coasts of Greenland (Hughes, 1986). However, there are probably negative feedback mechanisms that are not captured by our model and which could stabilize the ice sheet far short of a major collapse. If partial collapse of the Greenland Ice Sheet lowered the ice surface by 30 percent in 300 years, sea level would rise about 2 m and icebergs would flood the North Atlantic and produce a lid of cold fresh water as they melted (Hughes, 2004).

The Polar Front would move toward the Equator and push the North Atlantic Current toward Africa. The lid of melting icebergs would be semi-trapped in a gyre of cold fresh water north of the Polar Front that would suppress the ocean-to-atmosphere heat transport in the North Atlantic, so cold westerly winds would cause substantial cooling in northwest Europe not known since the Little Ice Age, from 1250 to 1850 in Greenland. The lid would suppress formation of North Atlantic Deep Water (NADW), which today forms in the Labrador Sea and north of Iceland. Here the "ocean conveyor belt" proposed by Broecker and Denton (1989) originates and brings NADW to the world ocean, providing our worldwide mild Holocene interglacial climate. If the lid of cold fresh water prevents NADW from forming, Little Ice Age conditions will prevail in western Europe. If the Greenland Ice Sheet continues to collapse, sea level could rise up to 5 m more and the new Little Ice Age could become a new 90,000-year worldwide glaciation cycle of the Quaternary Ice Age in which we have lived for the past million years.

## ACKNOWLEDGMENTS

Our work was supported by NSF and NASA through a grant from the Center for Remote Sensing of Ice Sheets (CReSIS) at The University of Kansas. We thank CReSIS Director, Prasad Gogineni, for this support and for providing the radar record of surface and bed topography down the centerline of Jakobshavn Isbrae from the interior ice divide to the grounding line of the terminal ice shelf in Jakobshavn Isfjord.

We also thank Joel Plummer for contributing to collecting these data. We thank Beverly Hughes for helping to prepare our manuscript, Olaf Eisen at the Alfred Wegener Institute for editorial input, Aitbala Sargent and James Fastook at The University of Maine for assistance with computer programming, and anonymous referees for sharpening our presentation.

## REFERENCES

Alley, R. B. (1989a), Water-pressure coupling of sliding and bed deformation: I. Water system, *Journal of Glaciology*, *35*(119), 108-118.

Alley, R. B. (1989b), Water-pressure coupling of sliding and bed deformation: II. Velocity-depth profiles, *Journal of Glaciology*, *35*(119), 119-129.

Anandakrishnan, S. and R. B. Alley (1997), Tidal Forcing of Basal Seismicity of ice stream C, West Antarctica Observed Far Inland, *Journal of Geophysical Research*, *102*(B7), 15,183–115,196.

Bassis, J.R. (2010). Hamilton's Principle applied to ice sheet dynamics: New approximations for the large-scale flow of ice sheets, *Journal of Glaciology*, 56(97), 497-513.

Bindschadler, R.A. (1984). Jakobshavn Glacier drainage basin: a balance asessment. *Journal of Geophysical Research*, 89, 2066-2072.

Bindschadler, R., and C. Hyengu (2007). Increased water storage in ice-stream onsets: a critical mechanism? *Journal of Glaciology*, 53(181), 163-171.

Brecher, H. H. (1986), Surface velocity determination on large polar glaciers by aerial photogrammetry, *Annals of Glaciology*, *8*, 22-26.

Broecker, W.S., and G.H. Denton (1989), The role of ocean-atmosphere reorganiztions in glacial cycles, *Geochimica et Cosmochimica Acta*, 53, 2465-2501.

Budd, W. F., D. Jenssen, and U. Radok (1971), Derived physical characteristics of the Antarctic Ice Sheet, Australian National Antarctic Research Expeditions (ANARE) Interim Reports, Series A (IV), *Glaciology*, 178 pages, University of Melbourne, Meteorology Department, Melbouorne, Australia.

Bueler, E., and J. Brown (2009), Shallow shelf approimation as a "sliding law" in a thermodynamically coupled ice sheet model, *Journal of Geophysical Research*, 114, F03008, doi: 10.1029/2008JF001179.

De Angelis, H. and P. Skvarca (2003), Glacier Surge After Ice Shelf Collapse, *Science*, *299*(Report), 1560–1562.

Denton, G. H., and T. J. Hughes (1981), *The Last Great Ice Sheets*, 484 pp., Wiley-Interscience, New York.

Dupont, T. and R. B. Alley (2005a), Assessment of the Importance of Ice-Shelf Buttressing to Ice-Sheet Flow, *Geophysical Research Letters*, *32*(4), LO4503.

Dupont, T. and R. B. Alley (2005b), The importance of small ice shelves in sea-level rise, *Geophysical Research Letters*, *33*, LO9503.

Echelmeyer, K., W. D. Harrison, T. S. Clarke and C. Benson 1992. Surficial glaciology of Jakobshavns Isbræ, West Greenland: Part II. Ablation, accumulation and temperature. *Journal of Glaciology*, *38*(128), 169-181.

Fastook, J. L., H. H. Brecher, and T. J. Hughes (1995), Derived bedrock elevations, strain rates, and stresses from measured surface elevations and velocities: Jakobshavn Glacier, Greenland. *Journal of Glaciology*, *41*(137), 161-173.

Glen, J. W. (1955), The creep of polycrystalline ice, *Proceedings of the Royal Society of London, Series A*(228), 519-538.

Gogineni, P. (2011), personal communication.

Hindmarsh, R. C. A. (2009), Consistent generation of ice-streams via thermo-viscous instabilities modulated by membrane stresses, *Geophysical Research Letters*, *36*(6), L06502. doi: 06510.01029/02008GL-036877.

Hindmarsh, R. C. A. (2012), An observationlly validated theory of viscous flow dynamics at the ice-shelf calving front. *Journal of Glaciology*, *58*(208), 1-13.

Hindmarsh, R. C. A., and E. LeMeur (2001), Dynamic processes involved in the retreat of marine ice sheets, *Journal of Glaciology*, *47*(157), 271–282.

Holland, D. M., R. H. Thomas, B. deYoung, M. H. Ribergaard, and B. Lyberth (2008), Acceleration of Jakobdshavn Isbrae triggered by warm subsurface ocean waters. *Nature Geoscience*, *1*(10), 659-664.

Hughes, T. (1986), The Jakobshavns Effect, *Geophysical Research Letters*, *13*(1), 46-48.

Hughes, T. (1992), On the pulling power of ice streams, *Journal of Glaciology*, *38*(128), 125-151.

Hughes, T. (1998), *Ice Sheets*, 343 pages, Oxford University Press, New York.

Hughes, T. (2009a), Modeling ice sheets from the bottom up, *Quaternary Science Reviews*, *28*(19-20), 1831-1849.

Hughes, T. (2009b), Variations in ice-bed coupling beneath and beyond ice streams: The force balance, *Journal of Geophysical Research*, *114*(B01410), doi:10.1029/2008JB005714.

Hughes, T. (2009c), Thermal Convection and the Origin of Ice Streams, *Journal of Glaciology*, *55*(191), 524-536.

Hughes, T. (2011), A simple holistic hypothesis for the self-destruction of ice sheets, *Quaternary Science Reviews*, *30*(15-16), 1829-1845.

Hughes, T. (2012), *Holistic Ice Sheet Modeling:* A First-Order Approach, 261 pages, Hauppauge, New York, Nova Science Publishers, Inc.

Hughes, T., A. Sargent, and J. L. Fastook (2011), Ice-Bed Coupling Beneath and Beyond Ice Streams: Byrd Glacier, Antarctica, *Journal of Geophysical Research*, *116*, 1-17.

Huybrechts, P. (1990), A 3-D model for the Antarctic Ice Sheet: a sensitivity study on the glacial-interglacial contrast, *Climate Dynamics*, *5*(2), 79-92.

Huybrechts, P. (1992), The Antarctic Ice Sheet and environmental change: a three-dimensional modelling study, *Ber. Polarforschung*, *99*, 241 pages.

Huybrechts, P. (1994), The present evolution of the Greenland Ice Sheet: an assessment by modelling, *Global and Planetary Change*, *9*, 39–51.

Huybrechts, P. (1996), Basal temperature conditions of the Greenland ice sheet during the glacial cycles, *Annals of Glaciology*, *23*, 226-236.

Iken, A., and M. Truffer (1997). The relatioship between subglacial water pressure and velocity of Findelengletscher, Switzerland, during its advance and retreat. *Journal of Glaciology*, 43(144), 328-338.

Iken, A., K. Echelmeyer, W. Harrison, and M. Funk (1993), Mechanics of fast flow in Jakobshavns Isbrae, West Greenland: Part I. Measurements of temperature and water level in deep boreholes, *Journal of Glaciology*, *39*(131), 15-25.

Jacobs, S. S., H. H. Hellmer, and A. Jenkins (1996), Antarctic Ice Sheet melting in the Southeast Pacific, *Geophysical Research Letters*, *23*(9), 957-960.

Johnson, J. (2002), A Basal Water Model for Ice Sheets, Ph.D. thesis, 187 pp, University of Maine, Orono.

Johnson, J., and J. L. Fastook (2002), Northern Hemisphere glaciation and its sensitivity to basal melt water, *Quaternary International*, *95–96*, 65–74.

Johnson, J., P. P. Prescott, and T. Hughes (2004), Ice dynamics preceding catastrophic disintegration of the floating part of Jakobshavn Isbrae, Greenland, *Journal of Glaciology*, *50*(171), 492–504.

Joughin, I., W. Abdalati, and M. Fahnestock (2004). Large fluctuations in speed on Greenland's Jakobshavn Isbrae glacier, *Nature*, 432, 608-610.

Joughin, I. R., I. M. Howat, M. A. Fahnestock, B. Smith, W. B. Krabill, R. B. Alley, H. Stern and M. Truffer (2008), Continued evolution of Jakobshavn Isbrae following its rapid speedup. *Journal of Geophysical Research*, *113*(F04006), doi:10.1029/2008JF001023.

Kamb, B. (1970), Sliding motion of glaciers: theory and observation, *Reviews of Geophysics and Space Physics*, *8*, 673-728.

Kamb, B. (2001), Basal Zone of the West Antarctic Ice Streams and its Role in Lubrication of Their Rapid Motion, in *The West Antarctic Ice Sheet: Behavior and Environment*, edited by R. B. Alley and R. A. Bindschadler, pp. 157–200, American Geophysical Union (Antarctic Research Series), Washington, DC.

Kenneally, J. P., and T. Hughes (2006), Calving giant icebergs: old principles, new applications, *Antarctic Science*, *18*(3), 409–419.

Lüthi, M., M. Funk, A. Iken, S. Gogineni, and M. Truffer (2002). Mechanisms of fast flow in Jakobshavn Isbrae, West Greenland, Part III: measurement of temperature and water level in deep boreholes, *Journal of Glaciology*, 46, 369-385.

Lüthi, M., M. Fahnestock, and M. Truffer (2008). Calving icebergs indicate a thick ice layer of temperate ice at the base of Jakobshavn Isbrae, Greenland, *Journal of Glaciology*, 55, 563-566.

MacAyeal, D. R. (1989), Large-scale ice flow over a viscous basal sediment: Theory and application to Ice Stream B, Antarctica, *Journal of Geophysical Research*, *94*(B4), 4071-4087.

MacAyeal, D. R., and R. H. Thomas (1982), Numerical modeling of ice-shelf motion, *Annals of Glaciology*, *3*, 189–194.

MacAyeal, D. R., T. A. Scambos, C. I. Hulbe, and M. A. Fahnestock (2003), Catastrophic Ice-Shelf Break-Up by an Ice-Shelf-Fragment-Capsize Mechanism, *Journal of Glaciology*, *49*(164), 22–36.

MacAyeal, D. R., E. A. Okal, R. C. Aster, and J. N. Bassis (2009), Seismic observations of glaciogenic oceanwaves (micro-sunamis) on icebergs and ice shelves *Journal of Glaciology*, *55*(190), 193-206.

Mahaffy, M. W. (1976), A three-dimensional numerical model of ice sheets: tests on the Barnes Ice Cap, Northwest Territories, *Journal of Geophysical Research*, *81*, 1059–1066.

Morland, L. W. (1987), Unconfined ice-shelf flow, in *Dynamics of the West Antarctic Ice Sheet*, edited by C. J. Van der Veen and J. Oerlemans, pp. 99–116, Reidel Publishing Company, Dordrecht.

Pelto, M.S., T.J. Hughes, and H. H. Brecher (1989), Equilibrium state of Jakobshavn Isbrae, West Greenland, *Annals of Glaciology*, 12, 127-131.

Podlech, S., and A. Weidick (2004), A catastrophic break-up of the front of Jakobshavn Isbrae, West Greenland, 2002/2003. *Journal of Glaciology*, 50, 153-154.

Prescott, P. R., J. P. Kenneally, and T. J. Hughes (2003), Relating crevassing to nonlinear strain in the floating part of Jakobshavns Isbrae, West Greenland, *Annals of Glaciology*, *36*, 149–156.

Reeh, N. (1982), A plasticity theory approach to the steady-state shape of a three-dimensional ice sheet., *Journal of Glaciology*, *28*(100), 431-455.

Reeh, N. (1984), Reconstruction of the glacial ice covers of Greenland and the Canadian Arctic islands by three-dimensional, perfectly plastic ice-sheet modeling, *Annals of Glaciology*, *5*, 115-121.

Reusch, D., and T.J. Hughes (2003). Surface "waves" on Byrd Glacier, Antarctica. *Antarctic Science*, 15(4), 547-555.

Rignot, E. J. (1998a), Fast recession of a West Antarctic glacier, *Science*, *281*, 549–551.

Rignot, E. J. (1998b), Radar interferometry detection of hinge-line migration on Rutford Ice Stream and Carlson Inlet, Antarctica, *Annals of Glaciology*, *27*, 25–32.

Rignot, E. J. (1998c), Hinge-line migration of Petermann Gletscher, north Greenland, detected using satellite-radar interferometry, *Journal of Glaciology*, *44*(148), 469-476.

Rignot, E. J. (2001a), InSAR studies of Pine Island and Thwaites Glaciers: A review of changes and possible causes, paper presented at West Antarctic Ice Sheet Initiative (WAIS), Eight Annual Workshop, Algonkian Meeting Center, Stirling, VA.

Rignot, E. J. (2001b), Evidence for rapid retreat and mass loss of Thwaites Glacier, West Antarctica, *Journal of Glaciology*, *47*(157), 213–222.

Rignot, E., and K. Steffen (2008), Channelized bottom melting and stability of floating ice shelves, *Geophysical Research Letters*, *35*(L02503), doi:10.1029/2007GL031765.

Rignot, E. J., S. Gogineni, I. Joughin, and W. Krabill (2001), Contribution to the glaciology of northern Greenland from satellite radar interferometry, *Journal of Geophysical Research*, *106*(D24), 34,007–034,019.

Rignot, E., J. Mouginot, and B. Scheuchl (2011), NASA research yields field map of Antarctic ice flow, in *NASA News*, edited, www.jpl.nasa.gov/news/news.cfm?release=2011-256 andcid=release.

Rist, M. A., P. R. Sammonds, S. A. F. Murrell, P. G. Meredith, H. Oerter and C. S. M. Doake (1996), Experimental fracture and mechanical properties of Antarctic ice: preliminary results. *Annals of Glaciology*, *23*, 85–102.

Rist, M. A., P.R. Sammonds, S.A.F. Murrell, P.G. Meredith, C.S.M. Doake, H. Oerter and K. Matsuki (1999), Experimental and theoretical fracture mechanics applied to Antarctic ice fracture and surface crevassing. *Journal of Geophysical Research*, *104*(B2), 2973–2987.

Rist, M. A., P. R. Sammonds, H. Oerter and C. S. M. Doake (2002), Fracture of Antarctic shelf ice. *Journal of Geophysical Research*, *107*(B1), ECV 1–13.

Robin, G. d. (1958), Glaciology III: Seismic shooting and related investigations, *Scientific Results of the Norwegian, British, Swedish Antarctic Expedition, 1949-1952*, *5*, 111-125.

Sargent, A. (2009), Modeling Ice Streams, Ph.D. thesis, 106 pp, University of Maine, Orono.

Schoof, C. (2007), Ice sheet grounding line dynamics: steady states, stability and hysteresis, *Journal of Geophysical Research Earth Surface*, *112*(FO3S28), doi: 10.1029/2006JF000664.

Schoof, C., and R. C. A. Hindmarsh (2010), Thin-film flows with wall slip: an asymptotic analysis of higher order glacier flow models, *Quarterly Journal of Mechanics and Applied Mathematics*, *63*(1), 73-114, doi:1093/qjmam/hbp1025.

Shepherd, A., D. J. Winkham, and, E. J. Rignot (2004), Warm ocean is eroding West Antarctic Ice Sheet, *Geophysical Research Letters*, *31*, L23402.

Sohn, H.-G., K. C. Jezek, and C. J. van der Veen (1998), Jakobshavn Glacier, West Greenland: 30 years of spaceborne observations, *Geophysical Research Letters*, *25*(14), 2699–2702.

Stearns, L. A., and G. S. Hamilton (2007), Rapid volume loss from two East Greenland outlet glaciers quantified using repeat stereo satellite imagery, *Geophysical Research Letters*, *34*(L05503), doi:10.1029/2006GL028982.

Sugden, D. E. (1977), Reconstruction of the morphology, dynamics, and thermal characteristics of the Laurentide Ice Sheet at its maximum, *Artic and Alpine Research*, *9*, 21-47.

Thomas, R. H. (1973a), The creep of ice shelves: theory, *Journal of Glaciology*, *12*, 45-53.

Thomas, R. H. (1973b), The creep of ice shelves: interpretation of observed behaviour, *Journal of Glaciology*, *12*, 55-70.

Thomas, R. H. (2004), Force-perturbation analysis of recent thinning and acceleration of Jakobshavns Isbrae, Greenland, *Journal of Glaciology*, *50*(168), 57-66.

Thomas, R. H., W. Abdalati, E. B. Frederick, W. B. Krabill, S. Manizade and K. Steffen (2003), Investigation of surface melting and dynamic thinning of Jakboshavns Isbrae, Greenland. *Journal of Glaciology*, *49*, 231–239.

Thomas, R.H., E. Frederick, J. Li, W. Krabill, S. Manizade, J. Paden, J. Sonntag, R. Swift and J. Yungel (2011), Accelerating ice loss from the fastest Greenland and Antarctic glaciers. *Geophysical Research Letters*, *3*, doi:10.1029/2011GL047304.

Weertman, J. (1957a), Deformation of floating ice shelves, *Journal of Glaciology*, *3*(62), 38-42.

Weertman, J. (1957b), On the sliding of glaciers, *Journal of Glaciology*, *3*(21), 33-38.

Weideck, A., H. Oerter, N. Reeh, H. H. Thomsen, and L. Thorning (1990), The recession of the Inland Ice margin during the Holocene climatic optimum in the Jakobshavns Isfjord area of West Greenland, *Palaeogeography, Palaeoclimatology, Palaeoecology (Global and Planetary Change Section)*, *82*, 389–399.

Weideck, A. and O. Bennike (2007), Quaternary glaciation history and glaciology of Jakobshavn Isbrae and the Disko Bugt region, West Greenland: a review, *Geological* Survey of Denmark and Greenland Bulletin 14, Ministry of Climate and Energy.

Zwally, H. J., W. Abdalati, T. Herring, K. Larson, J. Savba and K. Steffen (2002), Surface Melt-Induced Acceleration of Greenland Ice-Sheet Flow, *Science*, *297*(5579), 218–222.

In: Atmosphere and Climate
Editors: A. Wright and S. Johnson

ISBN: 978-1-62417-433-9


*Chapter 2*

# GLOBAL WARMING IN A CONTEXT OF THE MILLENNIAL CLIMATE HISTORY

## *M. Ogurtsov*[*]

Ioffe Physico-Technical Institute of Russian Academy of Sciences, St. Petersburg, Russia
Central Astronomical Observatory at Pulkovo, Russian Academy of Sciences, St. Petersburg, Russia

## ABSTRACT

Analysis of the available data on the Earth's global temperature over the past 150-1000 years is performed. Both instrumental (ground-based and satellite) data and proxies, obtained by paleoclimatic approaches, are analyzed. The estimation of advantages and shortcomings of various sources of the climatic information is made. Complex research of all the data set has shown that: (a) 20th century was actually warm, i.e. the global temperature of the Earth in 1900-2000 was above the millennial average; (b) in Northern hemisphere the time interval 1988-2008 likely was the warmest two decades over the last 1000 years. The potential origins of the considerable growth of temperature are considered. It is shown that the main factors likely responsible for a rise in global temperature are: (a) industrial emissions of greenhouse gases ($CO_2$, methane, nitrogen oxides); (b) natural climatic cycles; (c) solar-cosmic factors, including changes of solar luminosity, variation of the intensity of galactic cosmic

[*] Email: maxim.ogurtsov@mail.ioffe.ru.

ray and cosmic dust flux; (d) regional anthropogenic impact, including cutting down of woods, desertification, irrigation and ploughing of virgin lands. But it is difficult to estimate more or less precisely actual contribution of each factor due to our insufficient knowledge about them. It is shown that despite the successes and hard work in the field of both climatology and paleoclimatology our knowledge about the Earth's past climate is still not fully adequate. That is why it is hard to predict the change of the global climate in the current century. We can only consider some possible scenarios of the future climate evolution: (a) if global warming is a result of only industrial greenhouse effect, the rise of temperature will continue in $21^{th}$ century according to forecasts of IPCC; (b) if global warming is mainly a result of natural climatic cycles, in the first part of $21^{th}$ century the global temperature will decrease.

## INTRODUCTION

Climate has a long history as an important determinant of the social development of mankind. As a result, the well-known problem of *global warming* (GW) has already transferred from the field of pure scientific debates into a global political sphere. What is the physical mechanism of the GW? Is it resulted only from anthropogenic activity (especially burning of fossil fuels) or some other natural climatic phenomena may contribute to the global temperature increase?

Is the warming of the XX century really global? Is it actually unprecedent? The answer to these and many other questions can provide us valuable information about the climate behavior during at least few future decades and, hence, is of a great importance for all the humanity. E.g., if the climate was warmer at some point over the last millennium (the Medieval Warm Period or some other) then the lack of serious disasters during that period makes it possible to suggest that the further warming particularly would not be too destructive. Another major problem in global change studies is the detection of pure anthropogenic signal in climate, as it should be separated from the background of natural climatic variability. But in order to answer all the questions, stated above, one needs detailed information about the past of climatic variability and its causes. Just in this domain the modern knowledge has substantial gaps.

# GLOBAL WARMING OF 20TH CENTURY

The average temperature of the Earth, measured by the surface weather station thermometers, has appreciably grown during the last century. The IPCC consortium reported that the global mean surface temperatures have risen by 0.74°C ± 0.18°C when estimated by a linear trend over the last 100 years (IPCC, 2007). This rise is illustrated with figure 1A.

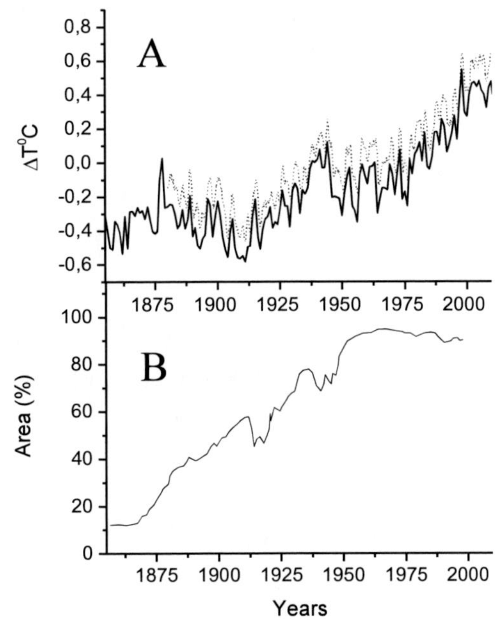


Figure 1.A. Observed changes in a global average temperature.
Solid line – http://www.cru.uea.ac.uk/cru/info/warming/;
dotted line –ftp://ftp.ncdc.noaa.gov/pub/data/anomalies/monthly.land_ocean.90S.90N.df_1901-2000mean.dat.
B. A part of the land covered with weather stations (Bell et al., 2003).

It is believed currently that the GW (a) is a mainly a result of anthropogenic greenhouse gases emission and (b) is unprecedent in a historical context. IPCC (2007) claims that the average Northern Hemisphere temperatures during the second half of the 20th century were higher than during any other 50-year period in the last 500 years with a probability > 0.9 and the highest in at least the past 1300 years with a probability > 0.66. But the historical instrumentally recorded temperatures exist only for 100 to 150 years. Moreover even available records are sparse and irregularly distributed. Modern weather station network covers less than 90% of the land i.e. no more than 25% of the Earth's surface – see Figure 1B. Satellite measurements of atmospheric temperature were begun in the end of 70s. They have much more dense spatial coverage. The satellite passes over most points on the Earth twice per day. Satellite-borne microwave sounders – the *microwave sounding units* (MSU) – estimate the temperature of thick layers of the atmosphere by measuring microwave emissions (radiances) that are proportional to the thermal state of emission of oxygen molecules from a complex of emission lines near 60 GHz. By making measurements at different frequencies near 60 GHz (≅ 1 cm), different atmospheric layers can be sampled.

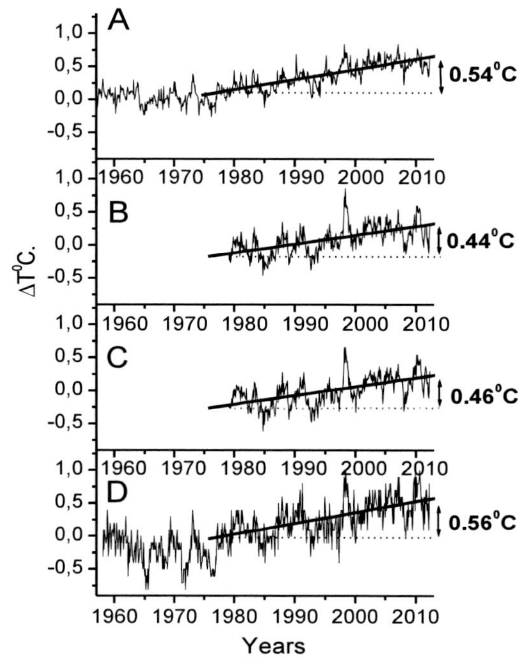


Figure 2.A. Observed changes in a global average temperature (ftp://ftp.ncdc.noaa.gov/pub/data/anomalies/monthly.land_ocean.90S.90N.df_1901-2000mean.dat) ;
B. RSS satellite based global lower troposphere temperature anomaly (http://www.remss.com/data/msu/);
C. UAH satellite based global lower troposphere temperature anomaly (http://vortex.nsstc.uah.edu/public/msu/);
D. Radiosonde global lower troposphere temperature anomaly MSU weighted (http://hadobs.metoffice.com/hadat/msu_equivalents.html).

Then, on the basis of the obtained data and by means of various mathematical procedures, the atmosphere temperature is calculated. Two groups of scientists – the Remote Sensing System (RSS) group and the University of Alabama (UAH) group – have analyzed the data of NASA satellites and obtained two versions of the change of temperature in the lower troposphere (heights less than 8 km, a maximum of sensitivity around 2-3 km) since the end of 1978. Figure 2 shows these two satellite series together with the surface thermometric data and measurements of the troposphere temperature (height about 1-8 km), made by radiosondes (weather balloons).

One can see from Figure 2 that the available data show that the temperature rise in the low troposphere is not stronger than at a surface as it should be in framework of greenhouse warming (see e.g. Santer et al. (2003), Douglass et al. (2004)). It should be noted however that satellite measurements have difficulties connected with the following problems: (a) offsets in calibration between satellites; (b) orbital decay and drift and associated long-term changes in the time of day that the measurements are made at a particular location; (c) drifts in satellite calibration. That is why it is not easy to evaluate correctly long-term trends of the satellite microwave-based data. Radiosonde data have even more limited spatial coverage than the surface thermometric records.

They usually are restricted with the middle-latitude land area. Thus study of instrumental (surface, satellite and radiosonde) data on the Earth's temperature change during the last 30-120 years is not enough to make a decisive conclusion about character and extent of the GW. That is why understanding of climate history in the pre-instrumental epoch is of a great importance. This knowledge could rely on climatic proxies – some physical data or measurement that can potentially serve as an indirect record of local temperature conditions.

## Methods of Studying of the Earth's Climate in the Past

Paleoclimatology is a science about a climatic history of the Earth. It reconstructs and studies climatic variations prior to the beginning of the instrumental era by means of different proxy indicators. Among the main data used by paleoclimatology are: tree rings (width and density), concentration of stable isotope ($^{18}O$, $^{13}C$, D) in natural archives (ice, coral and tree rings) and historical documents.

## Dendroclimatology

American astronomer A.E. Douglass is often considered as the founder of dendroclimatology – the scientific discipline, which estimates the past climate conditions from the dated tree rings. The basic idea of dendroclimatology is that rings of trees which grow under severe conditions react discernibly to variation of the limiting environmental factors. Thus trees living in extremely cold condition (northern tree line, upper tree line in mountains) reflect changes of summer temperature. Latewood density usually indicates temperature variation better than tree ring width. Tree rings are the important climate proxies because they can be absolutely dated annually. The works of Fritts (1976), Cook and Kairiukstis (1989) has helped dendroclimatology to become popular and over the last decades its methods have become key tools in reconstruction of past temperatures in many parts of the world (see e.g. Briffa, 2000; Mann and Hughes, 2002; Esper et al., 2002; Helama et al., 2005; Tuovinen et al., 2009) Tree-ring reconstructions usually are calibrated towards instrumental temperature. Coefficient of correlation between instrumental temperature series and tree-ring reconstructions reaches 0.4–0.5 over annual time scale and more than 0.8 over decadal time scale. For latewood density reconstructions these values are 0.5–0.8 and over 0.9 respectively. It should be noted that ability of dendroreconstructions to reflect the long-term (centennial and longer-scale) climate variations is often limited due to standardization– the detrending treatment of the growth effect of aging.

## Stable Isotope Climatology

H. Urey could be considered as the founder of isotope paleoclimatology. In 1948 he revealed that the oxygen isotope fractionation between calcium carbonate and water depends on temperature. Thus, Urey (1948) proposed that the isotopic composition of carbonates could be used as a paleothermometer. Later W. Dansgaard suggested the idea of using the isotopic composition of glacier ice as a climatic indicator (Dansgaard, 1954). Since isotopes with different masses have different rates of chemical reactions, natural geophysical and geochemical processes might cause isotope fractionation, which depends on temperature. As a consequence concentration of stable isotope $^{18}O$, $^{13}C$, D in natural archives – polar and mountain ice, bottom sediments, corals, tree ring cellulose – carries information about past temperature. Stable isotope records have apparent seasonal cycle which can be used for annual dating accuracy. Deeper into the ice the annual layers become thinner and finally become indistinguishable. Therefore, the dating of the deeper layers depends considerably on the flow model.

The isotopic composition of wood cellulose is another area of paleoisotope temperature reconstructions. Biological and physical processes determine stable isotope $^{13}C/^{12}C$ ratios in organic matter during photosynthetic uptake of $CO_2$ from the air. Strong response of $\delta^{13}C$ in northern Finland (Kessi) to midsummer temperature was found by Hilasvuori et al., (2009). Coefficients of correlation between the $\delta^{13}C$ record from Kessi and temperature, measured at the weather stations, is 0.68 at annual time scale and 0.81 at decadal time scale.

Ocean coral skeletal rings, or bands, also impart paleoclimatological information. Cooler temperatures as well as denser water salinity tend to cause coral to use heavier isotopes in its structure.

## Other Sources of Paleoclimatic Information

Borehole temperature measurement, historical data, melt percentages etc. are the other proxies used by paleoclimatology. Terrestrial deep soil temperature borehole profiles, measured in boreholes, can be used to obtain an estimate of ground surface temperature changes back in time. In a contrast with the other climate proxies (tree rings, corals, ice cores, and historical documentary records), the borehole records do not require calibration against independent surface temperature data and could be assessed as direct temperature estimates. As opposed to dendroreconstructions, borehole temperatures are able to yield information on long-term temperature trends only due to the specificity of the heat diffusion process in the soil column. Borehole data have been recently used to characterize Northern Hemisphere continental temperature for the past 500 years (Beltrami., 2002; Huang, 2004).

Contemporary written historical records – diaries, annals – contain information on a variety of natural phenomena: freeze dates, harvest amounts, flowering dates, extreme droughts, hurricanes etc. These data are often highly subjective and inhomogeneous. Thus their analysis is a complicated task.

Since different paleoindicators reflects actual temperature changes by different ways the *multiproxies* – the time series, which generalize proxy sets of various types – are often used by paleoclimatology.

## GLOBAL WARMING AND TREE-RING PROXIES

In Figure 3 seven reconstructions of the North Hemisphere temperature during the last millennium are shown. They are: dendroreconstructions by Briffa (2000) and Esper et al. (2002), non tree-ring reconstructions by Loehle (2007) and multi-proxies by Jones et al., (1998), Mann et al. (1999), Crowley and Lowery (2000), Moberg et al. (2005).

Despite multi-proxy reconstructions used many paleoindicators they are based mainly on tree-ring data. E.g. the reconstruction by Mann et al. (1999) was obtained using 12 proxy indicators – 7 tree-ring width based, 2 tree-ring density based, 2 based on $\delta^{18}O$ in ice core, 1 based on ice accumulation rate. Reconstruction by Jones et al. (1998) was obtained using 10 proxy indicators – 4 tree-ring width based, 1 tree-ring density based, 2 based on melt layers in ice cores, 1 based on direct temperature measurement, 1 based on historical data, 1 based on $\delta^{18}O$ in ice core.

One can see from Figure 3, that the millennial temperature reconstructions demonstrate visually different histories of temperature variability over the past 1000 years. They could be divided in the following groups:

a) Temperature reconstructions by Jones (1998), Mann (1999) and Crowley (2000) (Figure 3A,B,C) show apparent downward linear trend until the end of the 19$^{th}$ century and a sharp increase thereafter – the ice-hockey stick shape. These proxies show that Earth in the 20$^{th}$ century is substantially warmer than it has been in any other time period in the last millennium.
b) Reconstructions by Moberg (2005) and Loehle (2007) (Figure 3D,E) show a linear trend over ca AD 1000-1600. The warming of 20$^{th}$ century is comparable with that during the Medieval Warm Period (800 – 1100 AD) and represents a natural climate cycle with a period longer than 1000 year at least in part.
c) Reconstructions by Esper et al. (2002) and Briffa (2000) (Figure 3F,G) do not show a linear trend. Instead, their long-term variability is

dominated by multi-centennial variations. The 20th century was warm, but the warming is not anomalous.

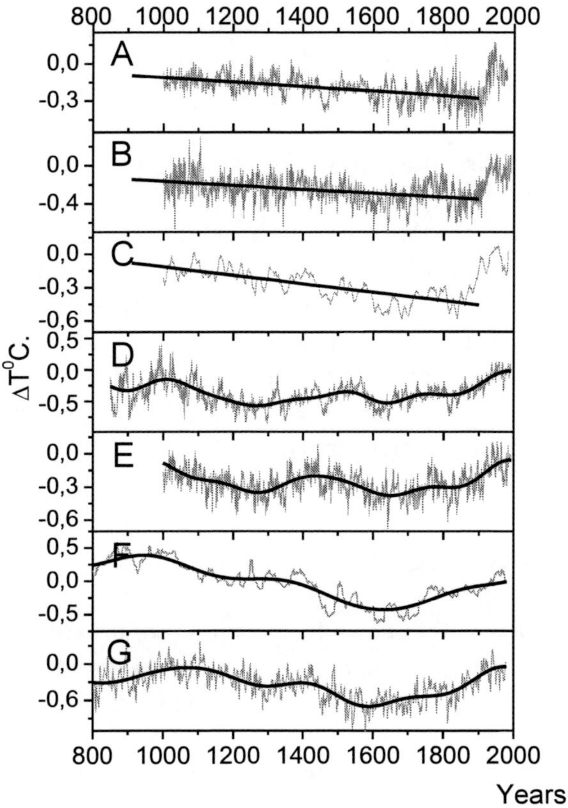


Figure 3. Reconstructions of the Northern Hemisphere temperature over the last millennium:
A. The multiproxy of Mann et al. (1999).
B. The multiproxy of Jones et al. (1998).
C. The multiproxy of Crowley and Lowery (2000).
D. The tree-ring proxy of Esper et al. (2002).
E. The tree-ring proxy of Briffa (2000).
F. The non tree-ring proxy of Loehle (2007).
G. The multiproxy of Moberg et al. (2005).
All the data sets, with the exception of the series after Loehle (2007) and Moberg (2005), were recalibrated by Briffa and Osborn (2002). Grey curves represent raw data. Thick black lines represent long-term trends.

It has been shown that the disagreement between the different temperature patterns cannot be fully explained by differences neither in standardization techniques nor in the different geographical coverage (Ogurtsov and Lindholm, 2006). That means that despite all the reconstructions show that the 20th century is warm and possibly one of the warmest of the last 10 centuries it is difficult to draw a plausible inference about the strength of the 20th century temperature rise (is it e.g. absolutely anomalous and unprecedented) before the problem of contradiction between climatic proxies is solved. This arises a problem of reliability of the available temperature reconstructions, which has become especially sharp after the works of Mc Intyre and Mc Kitrick (2003, 2005). These authors applied the methods of data transformation used by Mann et al. (1999) to the same source data and obtained the North Hemisphere temperature reconstruction rather different from the record of Mann et al. (1999). The works of Mc Intyre and Mc Kitrick (2003, 2005) have forced serious scientific debates (see Mann et al., 2003; von Storch et al. 2004). Existence of problems, which have not been solved till now, testifies that, despite evident successes, methods and approaches of paleoclimatology still leave considerable space to subjectivity.

## Global Warming and Borehole Data

The measurements of borehole temperature profiles are the only source of direct information about past temperatures. In contrast to dendrochronological proxies, geothermal records mainly reflect long-term temperature changes. Thus, although land use and vegetation cover changes may distort borehole temperature, these data play an important role in modern paleoclimatic research.

In Figure 4A,B,C, we have plotted three global surface temperature reconstructions – a series by Rutherford and Mann (2004), who analyzed 453 individual Northern Hemisphere borehole temperature records and a series by Beltrami (2002) made of 826 temperature-depth boreholes profiles distributed worldwide. These geothermal reconstructions are plotted together with the instrumental data.

The geothermal reconstructions, plotted in Figure 4, demonstrate that the 20th century (1900-2000 AD) is the warmest period over the last 500 years. Long-term trend of borehole-based estimations agrees well with that of directly measured temperature but doesn't follow well the carbon dioxide concentration. Really, as one can see from Figure 4c, rise of atmospheric $CO_2$ concentration started with the beginning of industrial revolution in the second part of 18th century, while global surface temperature increased since the beginning of 16th century. Thus borehole data shows that a 20th century warming is partly a result of very long climatic cycles not connected with human activity. Ogurtsov and Lindholm (2006) showed that the borehole-based reconstructions depict a history of past temperature changes conflicting with the ice-hockey stick proxies. The proxies by Mann et al. (1999), Jones et al. (1998), Crowley (2000) show a downward linear trend during 1500-1880 AD while the corresponding trend in the global borehole temperature is evidently upward. Other temperature reconstructions are in better agreement with the borehole data.

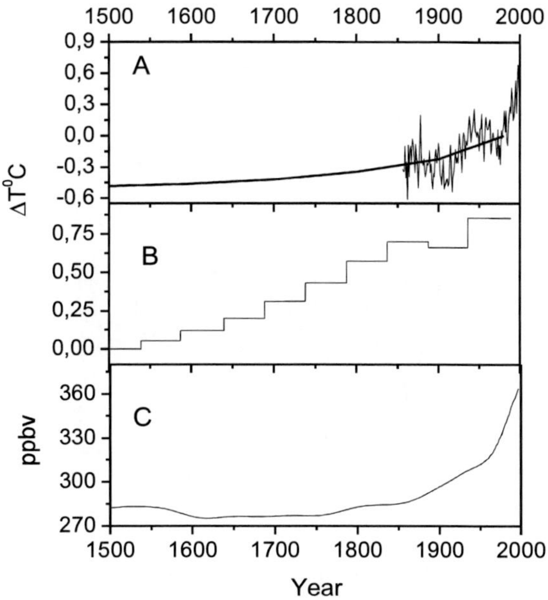


Figure 4. Borehole-based global temperature reconstructions.
A. Mann et al. (2004).
B. Beltrami (2002).
C. Concentration of carbon dioxide in the atmosphere (Robertson et al., 2001).

## Global Warming in Arctic

A separate analysis of other direct and indirect temperature data for locations situated far from human inhabitance may prove useful. The Arctic is a suitable region for this purpose. The Arctic is not only one of the least disturbed territories but also the region where greenhouse warming is expected to be clearly amplified according to global circulation models. Figure 8 after Lindzen (1997) show that over Arctic region warming should be more then twice stronger than over the globe. Direct and proxy records of Arctic temperature are shown in Figure 5.

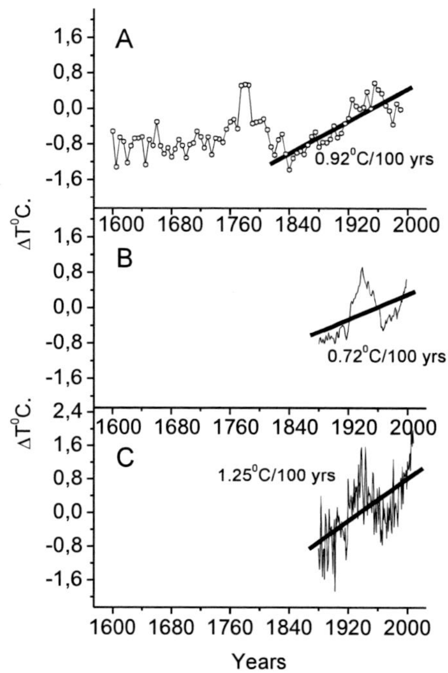


Figure 5. A. Reconstruction of the mean annual temperature in Arctic (Overpeck et al., 1997).
B. Mean annual temperature in Arctic measured instrumentally (Polyakov et al., 2003).
C. Mean annual temperature over latitudes higher than $65^0$N
(GHCN, http://www.co2science. org/ data/temperatures/).
Figures show value of the long-term trend.

Temperature record by Overpeck et al. (1997), who reconstructed annual temperatures in the Arctic using a number of proxy records (tree-rings, glacier data, and marine and lake sediments) is plotted in Figure 5A. Temperature record by Polyakov et al. (2003), who analyzed data from 75 arctic stations, is plotted in Figure5. Temperature data of Global Historical Climatology Network (GHCN) are plotted in Figure 5C. It is seen from Figure 5C that the highest Arctic temperatures have actually been recorded in the beginning of the current century. The linear warming trend in Arctic over the last 100-120 years is, however, less than $1.25^0C/100$ years i.e. only about 1.7 of the global temperature trend ($0.74^0C/100$ years). Another question – is a sharp rise of Arctic temperature connected with a greenhouse effect? Polar amplification of warming in general is a result of a positive feedback:

$$T^0C\uparrow \rightarrow \text{snow and ice area} \downarrow \rightarrow \text{albedo} \downarrow \rightarrow T^0C\uparrow \quad (1)$$

However, Graversen et al. (2008) have come to a conclusion that the maximum warming in Arctic regions occurs in "dark" months (November-February) when everywhere snow lies constantly and no changes in albedo occur.

Polyakov et al. (2005) concluded that warming of Arctic regions is a consequence of increase in temperature of warm Atlantic waters arriving in this zone. Polyakov et al. (2005) revealed appreciable warming of intermediate (depth 150-900) waters arriving in Arctic Ocean from Atlantic. This temperature anomaly extends from the west to the east from Barents sea to coast of Alaska (Polyakov et al. 2005).

## Climate Modeling and Forcing Factors

Climate modeling uses quantitative approaches to reproduce climatic change. Climate models are the sets of mathematical differential equations simulating the interactions of the atmosphere, mixed and deep ocean, land surface and cryosphere. Climate modeling has a variety of purposes from study of the dynamics of the climate system to prediction of future climate. The results of experiments with climate models are often considered as a confirmation of the reality of greenhouse-induced global warming (e.g. Crowley, 2000; Matthews et al., 2004; IPCC, 2007). The main drawback of such investigations is that none of the factors that may affect the Earth's climate are known precisely. Let us first consider solar activity, a factor effectively able to influence Earth's climate. Solar radiation is the only source of energy for the terrestrial atmosphere. The Sun's activity increased substantially during the $20^{th}$ century. Thus, solar activity seems to be a plausible candidate for producing a good part of GW. The most evident direct effect of solar variability on climate is its affect the Earth's radiative equilibrium through variations in luminosity. Total solar irradiance (TSI) is the agent which directly influences the terrestrial climate. What is more, Kärner

(2002) reported that TSI variations might play a dominant role in fluctuations in the lower troposphere temperature over a short-term (less than 20 years) time scale. Many TSI reconstructions covering the last few centuries have been obtained by means of different initial indicators ($^{10}$Be, $^{14}$C, sunspot number) – see e.g. Hoyt and Schatten (1993), Lean et al. (1995), Solanki and Fligge (1998), Beer et al. (2000), and Mordvinov et al. (2004). Nevertheless, our knowledge about the long-term changes of TSI is still poor. The existing solar proxies differ significantly in the amplitude of TSI variations. In Figure 6, some of the TSI records are shown.

Figure 6 shows that in the reconstruction by Hoyt and Schatten the increase in TSI since the end of the 19$^{th}$ century is about 3 Wt×m$^{-2}$, while in the record published by Mordvinov et al. (2004) it is only a bit more than 1 Wt×m$^{-2}$. After multiplication by 0.7 (taking the albedo into account) and dividing by 4 (averaging over the whole surface), these values lead to corresponding radiative forcings 0.52 Wt×m$^{-2}$ and 0.18 Wt×m$^{-2}$. Taking into account that characteristic climate sensitivities are 0.53-1.23 $^{0}$K×W$^{-1}$×m$^{2}$ (IPCC, 2007), the possible solar contribution to the warming during the 20$^{th}$ century can be evaluated as 0.09$^{0}$C (the lower values of the rise of the solar radiative forcing and climate sensitivity) to 0.64$^{0}$C (the upper values of rise of the solar radiative forcing and climate sensitivity). The difference between the lower and upper limits of the solar induced increase in the global temperature exceeds a factor 6. Such a great difference does not allow us to draw any decisive conclusion about the role of the Sun in global warming. It may be significant as well as negligible. According to IPCC (2007) the level of scientific understanding of solar contribution in the GW is low. Forcing caused by human-made changes in land surface properties since pre-agricultural times is unclear too. A possible value lies between 0.0 and 0.4 W×m$^{-2}$ (IPCC, 2007). Substantial uncertainty remains over the magnitude of influence of volcanic eruptions on the climate system. The system of feedbacks effectively controlling variations in the Earth's climate has not been finally established either (see Knutti et al., 2008).

Numerical experiment of Ogurtsov (2007) showed that an arbitrary picking of radiative forcings without justification of the choice criteria makes it possible to explain to a great extent the global warming of the 20$^{th}$ century beyond the hypothesis about the greenhouse effect. It is obvious that the parameters of many of the climate models have huge uncertainties thus it is difficult to assess their often quoted results as realistic or plausible.

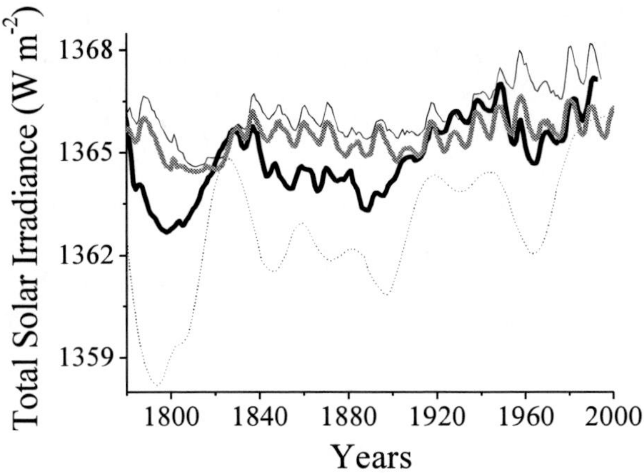


Figure 6. Reconstructions of total solar irradiance. The thick black line was obtained by Hoyt and Schatten (1993) and rescaled by Bard et al. (2000). The thin black line was obtained by Lean et al. (1995). The dotted black line was obtained by Beer et al. (2000). The thick grey line was obtained by Mordvinov et al. (2004). The data of Bard et al. (2000) and of Beer et al. (2000) were electronically scanned and digitized.

## FORECAST OF FUTURE CLIMATIC CHANGE

Instrumental and paleoclimatic the information available currently, allows consider two possible general scenarios of development of a global climate in the 21st century:

1) The observed GW is unique and unprecedented. It is connected, mainly, with the factor not operating before the beginning of the $20^{th}$ century i.e. the temperature history of last 1000 is compatible with ice-hockey stick reconstructions. In this case greenhouse warming due to industrial emission of greenhouse gases is the most plausible source of warming. Global warming appears as a result of industrial activity of mankind, and the past of climatic system can be neglected. Accordingly, forecasts of the evolution of the mean global temperature in the future could be found in works IPCC. These forecasts, which are made on a basis of various assumptions about the further development of industrial emissions of greenhouse gases, give temperature growth in the $21^{st}$century on 1.8-4.0 $^{0}$C (IPCC, 2007).

2) The observed GW is not unique in a historical context and is mainly the result of natural (terrestrial, solar, cosmophysical) climatic cycles i.e. the temperature history of last 1000 years is compatible with reconstructions by Esper et al. (2002), Briffa (2000), Moberg (2005) and Loehle (2007). In this case it is possible to suggest that the climate in the 20th century was driven by the same dynamic system, as during the last millenium. If it is actually so, we can predict future global temperatures using non ice-hockey stick millennial proxies as a source of information. Ogurtsov and Lindholm (2006) made a prognosis of mean decadal temperature over the first part of the 21st century by means of a nonlinear forecast technique (Hegger et al., 1999) and using reconstructions by Esper et al. (2002), Briffa (2000), Moberg (2005).The results are shown in Figure 7 together with the prediction error.

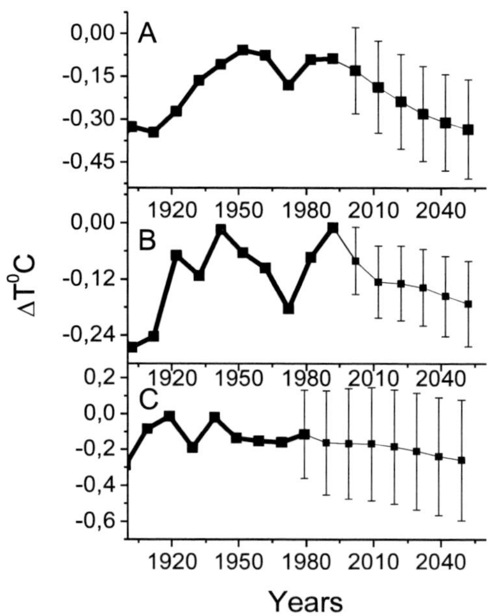


Figure 7. Forecasts of future global Northern Hemisphere temperatures (Ogurtsov and Lindholm, 2006).
A. Prediction based on data set Esper et al. (2002).
B. Prediction based on data set Briffa (2000).
C. Prediction based on data set Moberg (2005).
The original data, averaged by 13 years and interpolated by decades, are shown with thick black lines. Predicted values are shown with thin black lines.

Ogurtsov and Lindholm (2006) obtained that if the warming of the last 100 years is a result of natural climatic variability, i.e. the climate of past century is governed by the same dynamic system as the previous one to two millennia, the mean temperature of the first part of the $21^{st}$ century will unlikely be higher then the modern value.

If GW is a result of variety of climatic processes, each of which should be taken into account (anthropogenic emission of greenhouse gases (both industrial and agricultural), change of solar activity and space climate, internal fluctuations of climatic system, anthropogenic changes of a landscape) it is difficult to make any forecast of changes of a global climate because our knowledge on relative contributions of the specified factors is rather poor.

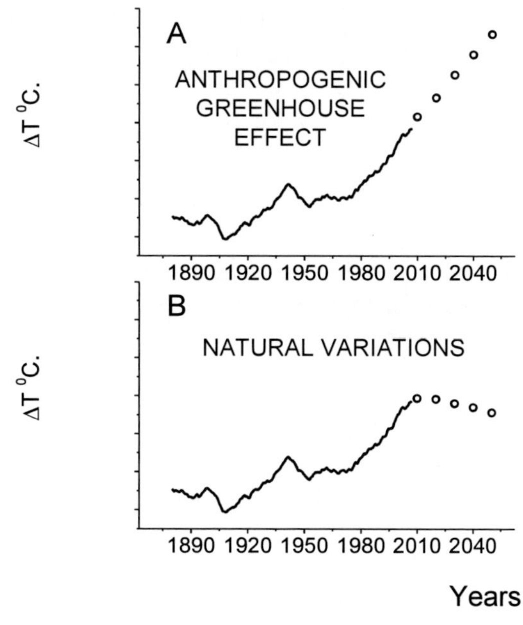


Figure 8. Possible qualitative scenarios of climatic evolution in the $21^{st}$ century.
A. If the GWof the $20^{th}$century has been caused by industrial greenhouse effect.
B. If the GW of $20^{th}$ century has been caused by the natural cycles of terrestrial, solar and cosmic nature.

## CONCLUSION

The analysis of the available information on the global temperature of the Earth, including both the data of instrumental measurements, and indirect paleoindicators, allow to conclude that the $20^{th}$ century was warm i.e. the average temperature of the Earth over AD 1900-2000 was higher than the average temperature through AD 1000-2000. The question, whether the temperature rise of the last century is unprecedented or similar warmings took place in the past, remains opened.

It is also difficult to estimate contribution of any of factors potentially responsible for the GW – industrial emission of greenhouse gases, growth of activity of the Sun, the regional anthropogenic impact, natural climatic cycles – due to insufficient information about the corresponding radiative forcings. The available information allows only to specify two possible scenarios of the global temperature evolution during the $21^{st}$ century (Figure 8) and to notice that if GW is caused, basically, by natural fluctuations of climatic system the average temperature of Northern Hemisphere of the Earth likely will not grow through the first part of $21^{st}$ century.

## ACKNOWLEDGMENTS

The author expresses his thanks to the exchange program between the Russian and Finnish Academies (project No. 16), to the program of Presidium of RAS No 22, and to RFBR grants No. 10-05-00129, 11-02-00755 for financial support.

## REFERENCES

Bard, E., Raisbeck, G., Yiou F., Jouzel J. (2000) Solar irradiance during the last 1200 years based on cosmogenic nuclides. *Tellus* Vol. 52B, pp. 985-992.

Beer, J., Mende, W. and Stellmacher, R. (2000). The Role of the Sun in Climate Forcing, *Quat. Sci. Rev.* Vol.19, pp. 403-415.

Bell J.L. et al. (2003). *Glob. and Planet. Change*. Vol. 37, pp. 19-32.

Beltrami, H. (2002). Climate from borehole data: energy fluxes and temperatures since 1500. *Geophys. Res. Let.* Vol.29, No.23, pp. 2111, doi:10.1029/2002GL015702.

Briffa, K.R. (2000). Annual climate variability in the Holocene: interpreting the message of ancient trees. *Quat. Sci. Rev.* Vol. 19, pp. 87-105.

Bürger, G. (2010). Clustering climate reconstructions. *Climate of the past. Discussions*. Vol. 6, pp. 659-679.

Crowley, T.J. and Lowery, T.S. (2000). How Warm Was the Medieval Warm Period? *Ambio*. Vol. 29, pp. 51-54.

Douglass, D.H., Pearson, B.D. and Singer S.F. (2004). Altitude dependence of atmospheric temperature trends: climate models versus observation. *Geophys. Res. Lett.* V. 31. DOI:10.1025/2004GL02103.

Esper, J., Cook E.R. and Schweingruber F.H. (2002) Low-frequency signals in long tree-ring chronologies for reconstructing past temperature variability. *Science*.Vol. 295, pp. 2250-2253.

Graversen, R.G., Mauritsen, T., Tjernstrom, M., Källén, E. and Svensson, G. (2008). Vertical structure of recent Arctic warming. *Nature*. Vol. 451, pp. 53-57.

Hegger, R., Kantz, H. and Schreiber, T. (1999) Practical implementation of nonlinear time series methods: the TISEAN package. *Chaos*. Vol. 9, pp. 413-435.

Hilasvuori, E.,Berninger, F., Sonninen, E., Tuomenvirta, H. and Jungner, H. (2009). Stability of climate signal in carbon and oxygen isotope records and ring width from Scots Pine (*Pinus sylvestris* L.) in Finland. *J. Quat. Sci.* Vol.24, No.5, pp. 469-480.

Helama, S., Timonen, M., Lindholm, M., MeriläInen, J. and Eronen, M. (2005). Extracting long-period climate fluctuations from tree-ring chronologies over timescales of centuries to millennia. *Int. J. of Climatology*. Vol. 25, pp.1767-1779.

Hoyt, D.V. and Schatten, K.H (1993) A discussion on plausible solar irradiance variations, 1700-1992. *J. Geophys. Res*. Vol. 98 (A11), pp. 18895-18906.

Huang, S. (2004). Merging information from different resources for new insights ino tclimate change in the past and future. *Geophys. Res. Lett.* Vol. 31, L13205, doi:10.1029/2004GL019781.

IPCC (2007). *WGI Fourth Assessment Report: Climate Change 2007: The Physical Science Basis: Summary for Policymakers,* Paris, 18 P.

Jones, P.D., Briffa, K.R., Barnett, T.P. and Tett, S.F.B. (1998). High-Resolution palaeoclimatic records for the last millennium: interpretation, integration and comparison with general circulation model control-run temperatures. *The Holocene*, Vol. 8.4, pp. 455-471.

Knutti, R., Allen, M. R., Friedlingstein, P., Gregory, J. M., Hegerl et al. (2008). A review of uncertainties in global temperature projections over the twenty-first century, *J. of Climate*. Vol. 21, pp. 2651-2663.

Lean, J., Beer, J. and Bradley, R. (1995). Reconstruction of Solar Irradiance since 1610: Implications for Climate Change. *Geophys. Res. Lett.* Vol. 22, No.23, pp. 3195-3198.

Lindzen, R.S. (1997). Can increasing carbon dioxide cause climate change? *Proc. Natl. Acad. Sci.* Vol. 94, No 16, pp. 8335-8342.

Ljungqvist, F.C. (2009). Temperature proxy records covering the last two millennia: a tabular and visual overview. *Geogr. Ann.*, Vol. 91A, pp. 11-29.

Loehle, C.A. (2007). 2000-year global temperature reconstruction based on non-treering proxies. *Energy and Envir.* Vol.18, No 7-8, pp.1049-1058.

Mann, M.E., Bradley, R.S. and Hughes, M.K. (1999). Northern Hemisphere temperatures during the past millennium: inferences, uncertainties, and limitations. *Geophys. Res. Lett.* Vol. 26, No.6,pp. 759-762.

Mann, M. and Hughes, M. (2002). Tree ring chronologies and climate variability. *Science*. Vol. 296, No.5569, pp. 848-852.

Mann, M.E., Zhang, Z., Hughes, M.K. et al. (2008). Proxy-based reconstructions of hemispheric and global surface temperature variations over the past two millennia. *Proc. Natl. Acad. Sci*. Vol. 105, No. 36, pp. 13252-13257.

Mann, M.E., Bradley, R.S. and Hughes, M.K. (2003). Note on Paper by McIntyre and McKitrick. In: Energy and Environment, *ftp://holocene.evsc.virginia.edu/pub/ mann/EandEPaperProblem.pdf*

Matthews, H.D., Weaver, A.J., Meissner, K.J. et al. (2004) Natural and anthropogenic climate change: incorporating historical land cover change, vegetation dynamics and the global carbon cycle. *Clim. Dyn.* Vol. 22, pp. 461–479.

Mc Intyre, S. and Mc Kitrick, R. (2003) Corrections to the Mann et al (1988) proxy data base and Northern Hemisphere average temperature series. *Energy Environ.* Vol. 14(6), pp. 751-772.

Mc Intyre, S. and Mc Kitrick, R. (2005) Hockey sticks, principal components and spurious significance. *Geophys. Res. Lett.* Vol. 32(3), doi:2004GL012750.

Moberg, A., Sonechkin, D.M., Holmgren, K. et al. (2005). High variable Northern Hemisphere temperatures reconstructed from low- and high-resolution proxy data. *Nature*. Vol. 433, pp. 613-617.

Mordvinov, A.V., Makarenko, N.G., Ogurtsov, M.G. and Jungner, H. (2004). Reconstruction of magnetic activity of the Sun and changes in its irradiance on a millennium timescale using neurocomputing. *Solar Physics*, Vol. 224, pp. 247-253.

Ogurtsov, M.G., Lindholm, M. (2006). Uncertainties in assessing global warming during the 20th century: disagreement between key data sources. *Energy Envir*. Vol. 17, pp. 685–706.

Ogurtsov, M.G., Jungner, H. and Lindholm, M. (2007). A potential century-scale rhythm in six major paleoclimatic records in the Northern Hemisphere. *Geograf. Ann.* Vol. 89A, pp. 129-136.

Ogurtsov M.G., Jungner H., Helama S., Lindholm M., Oinonen M. 2011: Paleoclimatological evidence for unprecedented recent temperature rise at the extratropical part of the Northern Hemisphere. *Geografiska Annaler*.V. 93, N1, P.17-27.

Polyakov, I.V., Beszczynska, A., Carmack E.C. et al.: (2005). One more step toward a warmer Arctic. *Geophys. Res. Lett.* Vol. 32, L17605, doi:10.1029/2005GL023740.

Robertson, A., Overpeck, J., Rind, D., et al. (2001) Hypothesized climate forcing time series for the last 500 years. *J. Geophys. Res.* Vol. 106(D14), pp. 14783-14803.

Santer, B.D., Wehner, M.F., Wigley, T.M.L., et al. (2003). Contributions of anthropogenic and natural forcing to recent tropopause height change. *Science*. Vol. 301, pp. 479–483.

Solanki, S.K. and Fligge, M. (1998). Solar irradiance since 1874 revisited. *Geophys. Res. Lett.* Vol. 30, pp. 333-351.

Tuovinen, M., Mc Carroll, D., Grudd, H., Jalkanen, R. and Los, S. (2009). Spatial and temporal stability of the climatic signal in Northern Fennoscandian pine tree-ring width and maximum density. *Boreas*, Vol.38, pp. 1-12.

Von Storch, H., E. Zorita, J.M., Jones P.D. et al. (2004). Reconstructing past climate from noisy data. *Science*. Vol. 306, pp. 679-682.

Urey, H.C. (1948). Oxygen isotopes in nature and in the laboratory. *Science*. Vol. 108, pp. 602-603.

In: Atmosphere and Climate
Editors: A. Wright and S. Johnson
ISBN: 978-1-62417-433-9


*Chapter 3*

# TEMPORAL AND SPATIAL PATTERNS OF $PM_{10}$ CONCENTRATIONS AT AN URBAN AREA

***J. C. M. Pires*** *

LEPAE, Departamento de Engenharia Química,
Faculdade de Engenharia, Universidade do Porto, Porto, Portugal

## ABSTRACT

Particulate matter is one of the most important air pollutants that negatively influence human health in Europe. They are emitted by both natural (volcanic eruptions, seismic activity and forest fires) and anthropogenic sources (all types of man-made combustion and some industrial processes). Long exposure to $PM_{10}$ and to $PM_{2.5}$ (particles with an aerodynamic diameter smaller than 10 and 2.5 μm, respectively) has been associated with respiratory and cardiovascular diseases. Air quality is strongly dependent on weather and, therefore, sensitive to climate change. Consequently, temporal and spatial patterns of air pollution are not only dependent on emission changes. The changes in the boundary layer height, relative humidity, temperature, solar radiation, wind speed and precipitation may be responsible for significant differences in air pollutant concentration patterns. Thus, this chapter aims to evaluate temporal and spatial patterns of $PM_{10}$ concentrations at an urban area

* Corresponding author: Telephone: +351 22 508 2262; Fax: +351 22 508 1449; E-mail address: jcpires@fe.up.pt.

through the application of principal component analysis (PCA). It was applied to air quality data collected in the Air Quality Monitoring Network of Oporto Metropolitan Area during the period from 2006 to 2010.

The monitored region presented high $PM_{10}$ concentrations (with several exceedances to European Union standard limits); however, they presented a decreasing tendency. PCA identified seven to nine $PM_{10}$ pollution patterns. Two monitoring sites often presented similar air pollution pattern. Analysing the annual average profile, $PM_{10}$ concentrations were higher during night time in winter period. This phenomenon is justified by the daily and seasonal variation of atmospheric boundary layer height, which influences the mixing of air pollutants.

**Keywords:** Air pollution, $PM_{10}$ concentrations, temporal tendency, principal component analysis

## 1. INTRODUCTION

Particulate matter (PM) is the term attributed to solid and liquid particles suspended in the atmosphere [1-3]. Some particles are large or dark enough to be seen as soot or smoke. Others are so small that only can be detected by electron microscopy. Its emissions sources can be natural or anthropogenic. The natural sources include volcanic eruptions, seismic activity and forest fires, while the main anthropogenic inputs are the combustion and some industrial processes [4-9]. Due to the variety of sources, their chemical and physical compositions vary widely. PM can be directly emitted (considered a primary pollutant) or can be formed in the atmosphere through reactions involving other air pollutants (considered a secondary pollutant). PM represents one of the most concerning air pollutant in Europe [10]. Their main consequences are the negative effects on human health and, in the last decades, several research studies were performed to associate its concentration with a wide range of human health problems [1, 11-14]. Long exposure to $PM_{10}$ and to $PM_{2.5}$ (particles with aerodynamic diameter smaller than 10 and 2.5 μm, respectively) has been associated with respiratory and cardiovascular diseases. For instance, research study performed by Kappos et al. [15] showed that PM is linked with elevated total, cardiovascular, and infant mortality. Aiming the protection of human health, the European Union (EU) established two limits for $PM_{10}$ concentrations: (i) the daily limit of 50 μg $m^{-3}$, not to be exceeded more than 35 times per year; and (ii) the annual limit of 40 μg $m^{-3}$ [16]. Regarding the influence on climate, PM modifies the energy balance of the Earth, the hydrological cycle, atmospheric circulation and the abundance of greenhouse gases [9, 17, 18]. PM can have positive or negative influence in radiative forcings, which represent changes in the energy fluxes of solar and terrestrial radiation in the atmosphere, induced by anthropogenic or natural changes in atmospheric composition, Earth surface properties or solar activity. Negative forcings include the scattering and reflection of solar radiation that tend to cool the Earth's surface, while positive forcings are associated with absorption of terrestrial radiation that tend to warm it (greenhouse effect). The effects are classified in direct and indirect with regards to radiative forcing of the climate system. Three physical mechanisms associated with PM effect on climate change are: (i) scattering and absorption of solar radiation; (ii) scattering, absorption and emission of thermal radiation; and (iii) changing cloud characteristics in many ways (for example, inhibition of freezing). The first two mechanisms are considered direct influences, while the last one is considered indirect effect [18-20]. As referred above, air pollutants have a strong influence on the Earth's climate system. Nevertheless, the climate change could also affect the air quality changing the chemical reaction rates, boundary layer heights (influence on vertical mixing of pollutants) and synoptic airflow patterns that control the transport of pollutants [21-23]. Consequently, air pollution is caused by interactions between local weather patterns, atmospheric circulation features, topography, human activities and its response to weather changes. As several phenomena associated to climate change are presenting higher frequency, temporal and spatial patterns of air pollution should be analysed to study the interaction between air pollution and climate change and to help the definition of policy decisions for mitigating these two environmental problems. Thus, this chapter aims to analyse temporal and spatial patterns of $PM_{10}$ concentrations at an urban area applying principal component analysis (PCA).

## 2. Data and Methods

### 2.1. Data

This study focused in the analysis of temporal and spatial variations of $PM_{10}$ concentrations collected in the Air Quality Monitoring Network (AQMN) of Oporto Metropolitan Area (Oporto-MA) from January 2006 to

December 2010. This region was selected as it presents high levels of $PM_{10}$ [22, 24, 25]. Its AQMN is managed by the Regional Commission of Coordination and Development of Northern Portugal (*Comissão de Coordenação e Desenvolvimento Regional do Norte*), under the responsibility of the Ministry of the Environment.

**Table 1. Site location and characteristics of the air quality monitoring network of Oporto Metropolitan Area (from www.qualar.org, accessed on August 2012)**

| Site | Latitude | Longitude | Altitude (m a.s.l.) | Type | Sampling period |
|---|---|---|---|---|---|
| AN | 41 09'52" N | 8 35'22" W | 146 | Urban - Traffic | 2000 – |
| BV | 41 09'33" N | 8 37'45" W | 87 | Urban - Traffic | 2006 – 2010 |
| CS | 41 11'59" N | 8 38'41" W | 100 | Suburban - Background | 1999 – |
| ER | 41 12'24" N | 8 33'10" W | 140 | Urban - Background | 1999 – |
| ES | 41 00'41" N | 8 38'18" W | 23 | Urban - Traffic | 2000 – 2010 |
| LB | 41 13'05" N | 8°37'56" W | 40 | Suburban - Background | 2000 – |
| MN | 41 18'31" N | 8 43'15" W | 25 | Suburban - Background | 2010 – |
| MT | 41 11'10" N | 8 40'42" W | 32 | Urban - Traffic | 2002 – 2010 |
| PR | 41 13'56" N | 8 42'47" W | 25 | Suburban - Industrial | 2002 – |
| SB | 41 08'51" N | 8 39'32" W | n.a. | Urban - Background | 2008 – |
| SH | 41 11'10" N | 8 39'47" W | 72 | Urban - Traffic | 2001 – |
| VC | 41 20'42" N | 8 44'14" W | 25 | Suburban - Traffic | 2002 – 2009 |
| VR | 41 14'08" N | 8 37'07" W | 90 | Urban - Traffic | 2001 – |
| VT | 41 15'08" N | 8 39'38" W | 88 | Suburban - Background | 1999 – |

n.a. – not available.

The most important air pollution sources in Oporto-MA are vehicle traffic, an oil refinery, a petrochemical complex, a thermoelectric plant (planned for working with natural gas), an incineration unit and an international shipping port [25, 26]. The AQMN of Oporto-MA is currently composed of 11 monitoring sites that regularly monitor the air levels of carbon monoxide, nitrogen oxides, sulphur dioxide, ozone and $PM_{10}$. $PM_{2.5}$, benzene, toluene and xylene pollutants are also measured in some monitoring sites of the network.

One monitoring site (*Avintes*) only has few measurements in 2010 and, consequently, it was removed from this study. However, four excluded monitoring sites were included for this temporal and spatial analysis of $PM_{10}$ concentrations. Thus, for the analysed period, $PM_{10}$ concentrations data is available for 14 monitoring sites: *Antas* (AN), *Boavista* (BV), *Custóias* (CS), *Ermesinde* (ER), *Espinho* (ES), *Leça do Balio* (LB), *Mindelo* (MN), *Matosinhos* (MT), *Perafita* (PR), *Sobreiras* (SB), *Senhora da Hora* (SH), *Vila do Conde* (VC), *Vermoin* (VR) and *Vila Nova da Telha* (VT). Table 1 shows the location and the main AQMN monitoring site characteristics, and the period of the available $PM_{10}$ measurements (from www.qualar.org, accessed on August 2012).

$PM_{10}$ concentrations were obtained through the beta radiation attenuation method, considered equivalent to the method advised by the EU Directive 2008/50/EC [16], using the MPSI 100 I et E equipment from Environment SA [25]. This equipment was submitted to a rigid maintenance programme, being calibrated each 4 weeks. Measurements were continuously registered and hourly average mass concentrations (in $\mu g\ m^{-3}$) were recorded.

## 2.2. Principal Component Analysis

PCA is mathematically defined as an orthogonal linear transformation that modifies the original data to a new coordinate system such that the greatest variance by any projection of the data comes to lie on the first coordinate (called the first principal component), the second greatest variance on the second coordinate, and so on [25, 26]. Thus, the principal components (PCs) are orthogonal and uncorrelated to each other, being determined by linear combinations of the original variables. The directions of the new coordinate axes are given by the eigenvectors of the covariance matrix of the original variables. The magnitude of the eigenvalues corresponds to the variance of the data along the eigenvector direction.

PCs are dependent on the units used to measure the original variables as well as on the range of values they assume. Thus, the data should be standardized before applying PCA. A common standardization method is to transform all the data to have zero mean and unit standard deviation. After the standardization of the original data, the covariance matrix is determined. The eigenvalues are calculated from the following equation [27]:

$$|\mathbf{C} - \boldsymbol{\lambda}\mathbf{I}| = \mathbf{0} \tag{1}$$

where C is the covariance matrix, λ are the eigenvalues, I the identity matrix and |...| is the matrix determinant operator. The eigenvectors are calculated by:

$$(\mathbf{C} - \boldsymbol{\lambda}\mathbf{I})\mathbf{W} = \mathbf{0} \tag{2}$$

where W is the matrix of the weights (or eigenvectors). The PCs are then obtained, multiplying the original data set by the weights.

Varimax rotation is the most widely employed orthogonal rotation in PCA, because it tends to produce simplification of the unrotated loadings to easier interpretation of the results. It simplifies the loadings by rigidly rotating the PC axes such that the variable projections (loadings) on each PC tend to be high or low. The interpretation simplicity of a specific PC is defined as the variance of its squared loadings [28]:

$$s_p^2 = \frac{1}{n}\sum_{j=1}^{n}\left(b_{jp}^2\right)^2 - \frac{1}{n^2}\left(\sum_{j=1}^{n} b_{jp}^2\right)^2 \tag{3}$$

where $n$ is the number of original variables and $b_{jp}$ is the loading of the original variable $j$ in the principal component $p$. If the variance is at maximum, the PC has the greatest interpretability. In this case, the loadings tend toward unit and zero. To rotate the loadings of all PCs, the sum of the variances of the squared loadings must be maximized:

$$s^2 = \sum_{p=1}^{m} s_p^2 = \frac{1}{n}\sum_{p=1}^{m}\sum_{j=1}^{n} b_{jp}^4 - \frac{1}{n^2}\sum_{p=1}^{m}\left(\sum_{j=1}^{n} b_{jp}^2\right)^2 \tag{4}$$

where $m$ is the number of PCs. The procedure to maximize the variance is described in more detail in [28]. After the rotation, the loadings show the relative contributions of the original variables on each PC.

**Table 2. Statistics of the hourly average $PM_{10}$ concentrations (in μg $m^{-3}$) collected in the air quality monitoring network of Oporto Metropolitan Area during the sampling period**

| Site | 2006 | | | 2007 | | | 2008 | | | 2009 | | | 2010 | | |
|---|---|---|---|---|---|---|---|---|---|---|---|---|---|---|---|
| | min | av | max | min | av | max | min | av | max | min | av | max | min | av | max |
| AN | 0 | 36 | 142 | 0 | 35 | 152 | 0 | 26 | 171 | 0 | 26 | 591 | 0 | 28 | 189 |
| BV | 0 | 39 | 235 | 0 | 37 | 192 | 0 | 28 | 204 | 0 | 27 | 164 | 0 | 24 | 119 |
| CS | 0 | 38 | 250 | 0 | 35 | 240 | 0 | 23 | 197 | 0 | 25 | 291 | 0 | 28 | 253 |
| ER | 2 | 40 | 199 | 1 | 39 | 178 | 0 | 29 | 183 | 0 | 30 | 150 | 0 | 29 | 164 |
| ES | 0 | 43 | 327 | 0 | 44 | 290 | 0 | 37 | 458 | 0 | 40 | 402 | 0 | 39 | 304 |
| LB | 0 | 38 | 320 | 0 | 38 | 410 | 0 | 42 | 283 | 0 | 29 | 244 | 0 | 26 | 216 |
| MN | | | | | | | | | | | | | 0 | 26 | 146 |
| MT | 0 | 41 | 276 | 0 | 42 | 235 | 0 | 29 | 232 | 0 | 28 | 144 | 0 | 32 | 170 |
| PR | 0 | 36 | 215 | 0 | 37 | 377 | 0 | 31 | 222 | 0 | 32 | 220 | 0 | 30 | 354 |
| SB | | | | | | | | | | 0 | 25 | 227 | 0 | 25 | 204 |
| SH | 0 | 40 | 260 | 1 | 40 | 197 | 0 | 29 | 221 | 0 | 33 | 138 | 0 | 40 | 769 |
| VC | 0 | 46 | 345 | 0 | 43 | 245 | 0 | 40 | 286 | 0 | 37 | 266 | | | |
| VR | 0 | 38 | 306 | 0 | 39 | 337 | 0 | 26 | 253 | 0 | 26 | 232 | 0 | 25 | 212 |
| VT | 0 | 33 | 531 | 0 | 30 | 135 | 0 | 26 | 134 | 0 | 24 | 164 | 0 | 46 | 281 |

min – minimum; av – average; max – maximum.

In environmental research studies, PCA is often applied as a classification method, identifying patterns on complex data matrixes. Several studies can be found that applied this statistical methodology for the management of the air quality monitoring networks [29-33]. Also many studies can be found concerning several air pollutants in AQMN of different cities: Santiago de Chile [34]; Porto [24-26]; Lisbon [35]; London [36]; Bilbao [37]; Beirut [38]; Hong Kong [39, 40]; and Beijing [41]. For instance, in the study performed for Oporto PCA identified similar patterns of air pollution in several monitoring sites (more than half). Consequently, it was proposed the removal or replacement of monitoring sites corresponding to the redundant measurements. After the selection of the monitoring sites, the air pollutant concentrations at the removed monitoring sites were well estimated by statistical models using the data collected at the remaining sites. Thus, the removed monitoring sites did not represent a significant loss of air quality information. The other studies showed also a significant reduction of the monitoring sites after the PCA application, concluding that this statistical methodology is also an important statistical tool for the management of the air quality monitoring networks.

# 3. Results and Discussion

## 3.1. Statistics and Exceedances to EU Limits

Table 2 shows the annual minima, average and maxima of hourly average $PM_{10}$ concentrations during the analysed period at all monitoring sites. These values show the spatial and also temporal variation of the concentrations of this air pollutant in the monitored region. Generally, $PM_{10}$ concentrations presented a decreasing tendency. The highest concentrations were measured at ES and VC sites. Pires et al. [25] had analysed the influence of wind direction on $PM_{10}$ concentrations in Oporto-MA. An emission source of this pollutant was located at E-SE of the monitored region (influencing the pollution pattern of all region) and another source located at NW-N of VC that strongly influences the concentrations measured at this site. In the same study, VC site also presented the highest concentrations (measuring period: 2003-2005). Figure 1 shows the distribution of $PM_{10}$ concentrations at ES and VC sites, presenting the monthly minima, lower quartile, median, upper quartile and maxima. A temporal pattern was not found for both sites. Contrary to surface ozone, the weather has lower impact in $PM_{10}$ concentrations. Having several and "seasonal" emission sources (industry and traffic during all year; heating during the winter; forest fires during summer), it is difficult to predict its concentrations. However, analysing the values presented in Figure 1, ES site presented more than 50% of the measured concentrations above 50 $\mu g\ m^{-3}$ (standard value defined for daily average concentrations for human health protection) during February 2006 and February 2009, while at VC site this occurs during February 2006, January 2007, January 2008 and February 2009. Similar analysis was also performed for the remaining sites and it is concluded that the highest concentrations generally occurred in winter periods. Only SH and VT sites presented the highest concentrations in summer (August 2010).

Table 3 shows the number of exceedances of the limit of 50 $\mu g\ m^{-3}$ established by the EU for human health protection and the correspondent monitoring efficiency, regarding the daily average concentrations of $PM_{10}$. In order to be in compliance, the limit cannot be exceeded more than 35 times a year which corresponds to 10% of the total number of available days [16]. The number of exceedances allowed by the EU was surpassed at almost all monitoring sites during the analysed years, with exception of MN and SB sites. It was also observed that there was a significant reduction of the number of exceedances in the last three years. During this period, AN, BV, CS and VR sites presented the number of exceedances below the allowed by the legislation. ES and VC sites presented the highest number of exceedances, representing about 30% of the available data. Other standard limit for human health protection is the annual average $PM_{10}$ concentration [16], which values for each monitoring site were presented in Table 2. The highest annual averages were generally observed in 2006 and 2007, period in which ES, MT and VC sites did not obeyed the correspondent standard limit (40 $\mu g\ m^{-3}$). On the other hand, LB and VT sites were not in compliance of the annual limit during 2008 and 2010, respectively. Besides the traffic and industrial emissions of $PM_{10}$, the regional transport played an important role in the pollutant concentration. The research made by Pereira et al. [42] showed that even at rural sites (absence of direct influence of emission sources) exceedances were observed due to intra-regional pollutant transport.

**Table 3. Number of exceedances of the limit established by the European Union for human health protection regarding daily average $PM_{10}$ concentrations (50 $\mu g\ m^{-3}$) and the correspondent monitoring efficiency (in brackets)**

| Site | 2006 | 2007 | 2008 | 2009 | 2010 |
|---|---|---|---|---|---|
| AN | 71 (98) | 62 (100) | 16 (96) | 21 (80) | 28 (100) |
| BV | 79 (92) | 70 (99) | 31 (93) | 27 (99) | 1 (46) |
| CS | 59 (63) | 73 (99) | 13 (90) | 32 (100) | 33 (99) |
| ER | 89 (90) | 86 (96) | 33 (99) | 39 (96) | 34 (100) |
| ES | 86 (85) | 130 (99) | 70 (98) | 93 (99) | 73 (87) |
| LB | 39 (44) | 75 (86) | 34 (35) | 39 (77) | 15 (63) |
| MN | | | | | 24 (92) |
| MT | 104 (98) | 116 (96) | 30 (87) | 28 (92) | 37 (83) |
| PR | 62 (94) | 67 (97) | 40 (98) | 46 (96) | 40 (97) |
| SB | | | | 12 (72) | 20 (95) |
| SH | 80 (82) | 92 (95) | 29 (77) | 46 (78) | 71 (96) |
| VC | 111 (95) | 115 (99) | 83 (98) | 59 (72) | |
| VR | 86 (93) | 89 (90) | 29 (93) | 35 (98) | 28 (96) |
| VT | 56 (100) | 36 (78) | 17 (80) | 31 (88) | 126 (99) |

The exceedances should not be more than 35 times a year.

### 3.2. PCA Results

To be able to apply the PCA, hourly average concentrations of $PM_{10}$ need to be available at all sites in the same time period. The row's data should be removed, when at least one value is missing. To avoid the lost of a significant fraction of data, the sites with low monitoring efficiency (below 75%) were removed from the annual analysis.

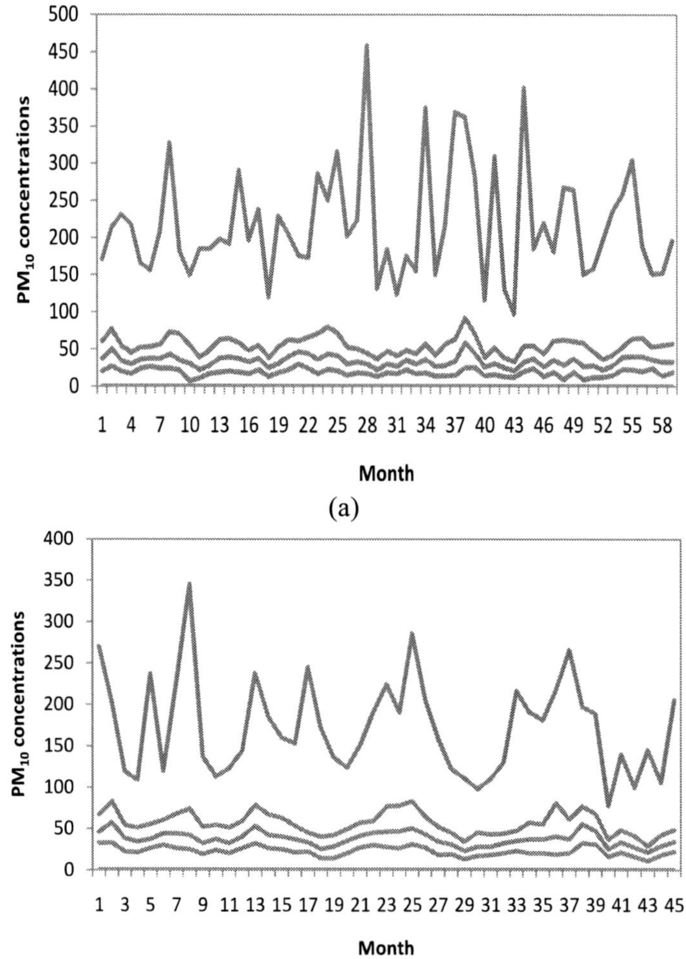


Figure 1. Temporal variation of the minima (always zero), lower quartile, median, upper quartile and maxima hourly average $PM_{10}$ concentrations (in μg $m^{-3}$) for each month measured at ES and VC (with less than 4 years of $PM_{10}$ measurements) sites.

Thus, the monitoring sites not considered in PCA were: (i) CS, LB, MN and SB in 2006; (ii) MN and SB in 2007; (iii) LB, MN and SB in 2008; (iv) MN, SB and VC in 2009; and (v) BV, LB and VC in 2010. After removing the missing data, the sizes of hourly data were 4799, 5121, 4179, 3350 and 5946 for the 2006, 2007, 2008, 2009 and 2010 years, respectively. The concentration values were Z standardized to have zero mean and unit standard deviation.

PCA was applied as a classification method, grouping the monitoring sites according their $PM_{10}$ pollution pattern. In other words, monitoring sites associated to a specific PC presented the similar variation of $PM_{10}$ concentrations and different variation from those belonging to other PCs. Table 4 shows, as an example, the main results of PCA application to $PM_{10}$ data collected in 2007, period corresponding to more monitoring sites [12]. The first step is the selection of the number of PCs.

The most common one is the Kaiser criterion [43] that selects the PCs with eigenvalues greater than 1. In this study, the application of this method would result in the selection of only one PC, representing 66.8% of original data variance. Other PCs should be selected to increase the explained variance, improving at same time the confidence in this statistical analysis. Considering at least 95% of the original data variance, 9 PCs were selected for the data of 2007 (the selected PCs were between 7 and 9 for the data of other years). Besides the cumulative variance of each PC, Table 4 presents also their factor loadings, which represents the correlation between the PCs and the original variables.

Higher absolute factor loadings (close to one) correspond to stronger relationships between PCs and the original variables. Thus, the first PC (PC1) had important contributions of AN, ER and SH sites, while PC5 was heavily loaded by LB and VR sites. BV, CS, ES, MT, PR, VC and VT sites were significantly associated with PC9, PC7, PC2, PC4, PC6, PC3 and PC8, respectively.

In other periods, AN and ER were frequently grouped, having redundant $PM_{10}$ measurements in almost all period (with exception of 2009). The behaviour of $PM_{10}$ concentrations at these sites was also similar to those observed at SH (2007 and 2008), VR (2006) and VT (2008 and 2010) sites. Similar $PM_{10}$ behaviour observed at LB and VR sites was also observed in previous studies [25, 26].

These two sites always presented different air pollution behaviour, regarding the five studied pollutants (sulphur dioxide, $PM_{10}$, carbon monoxide, nitrogen oxides and surface ozone). The geographical location of the main emission sources and the topography (they are located between two important valleys – from Leça river and from the Almorode river) were considered the important causes for this differentiation.

### 3.3. $PM_{10}$ Annual Profiles

$PM_{10}$ concentrations do not have a defined temporal pattern, being influenced by several emission sources and meteorological conditions. Figure 2 shows the annual average profile of $PM_{10}$ concentrations measured at 12 monitoring sites (excluding MN and SB due to few data) during the studied period. High concentrations were observed at all sites during night time periods of January, February, November and December. This phenomenon may be related with atmospheric physics, in particular the daily variation of the atmospheric boundary layer (ABL) stability. During the day, the sun heats the Earth's surface, uprising air masses (44). A significant part of ABL is in an adiabatic state, which defines the mixing layer (or convective layer). In this period, the convection of air masses is stronger, increasing the ABL height. On the other hand, during the night, the temperature inversion occurs at the ground, due to the cooling of the soil (it has a better thermal conductivity than air). This results in a formation of a stable layer (with lower height) in which air pollutant concentrations are increasing. Above this layer, the mixing layer of the previous day becomes a neutral residual layer, containing the pollutants that have been mixed the day before. Besides the cold periods, the average concentrations were also high in August and September, when $PM_{10}$ emissions from forest fires have more impact.

**Table 4. Main results of the PCA application (rotated factor loadings and cumulative variance) to $PM_{10}$ data collected in 2007**

| Site | PC1 | PC2 | PC3 | PC4 | PC5 | PC6 | PC7 | PC8 | PC9 |
|---|---|---|---|---|---|---|---|---|---|
| AN | -0.819 | -0.160 | -0.181 | -0.209 | 0.200 | -0.117 | -0.176 | -0.155 | 0.231 |
| BV | -0.459 | -0.185 | -0.191 | -0.208 | 0.233 | -0.189 | -0.263 | -0.175 | 0.681 |
| CS | -0.330 | -0.172 | -0.177 | -0.232 | 0.216 | -0.121 | -0.792 | -0.168 | 0.187 |
| ER | -0.834 | -0.142 | -0.154 | -0.133 | 0.279 | -0.156 | -0.176 | -0.180 | 0.110 |
| ES | -0.186 | -0.897 | -0.166 | -0.163 | 0.131 | -0.223 | -0.128 | -0.084 | 0.102 |
| LB | -0.533 | -0.067 | -0.150 | -0.121 | 0.586 | -0.242 | -0.381 | -0.108 | 0.028 |
| MT | -0.324 | -0.230 | -0.199 | -0.791 | 0.178 | -0.191 | -0.240 | -0.157 | 0.152 |
| PR | -0.228 | -0.280 | -0.259 | -0.166 | 0.154 | -0.831 | -0.122 | -0.145 | 0.123 |
| SH | -0.696 | -0.190 | -0.196 | -0.260 | 0.212 | -0.215 | -0.258 | -0.254 | 0.203 |
| VC | -0.249 | -0.194 | -0.851 | -0.168 | 0.159 | -0.248 | -0.160 | -0.142 | 0.121 |
| VR | -0.409 | -0.222 | -0.180 | -0.190 | 0.713 | -0.118 | -0.153 | -0.218 | 0.242 |
| VT | -0.454 | -0.131 | -0.206 | -0.192 | 0.241 | -0.203 | -0.215 | -0.722 | 0.157 |
| CV | 66.8 | 74.6 | 78.8 | 82.7 | 86.0 | 88.8 | 91.3 | 93.7 | 96.0 |

CV – Cumulative variance (%); values in bold indicate the variables that mostly influence the correspondent principal component.

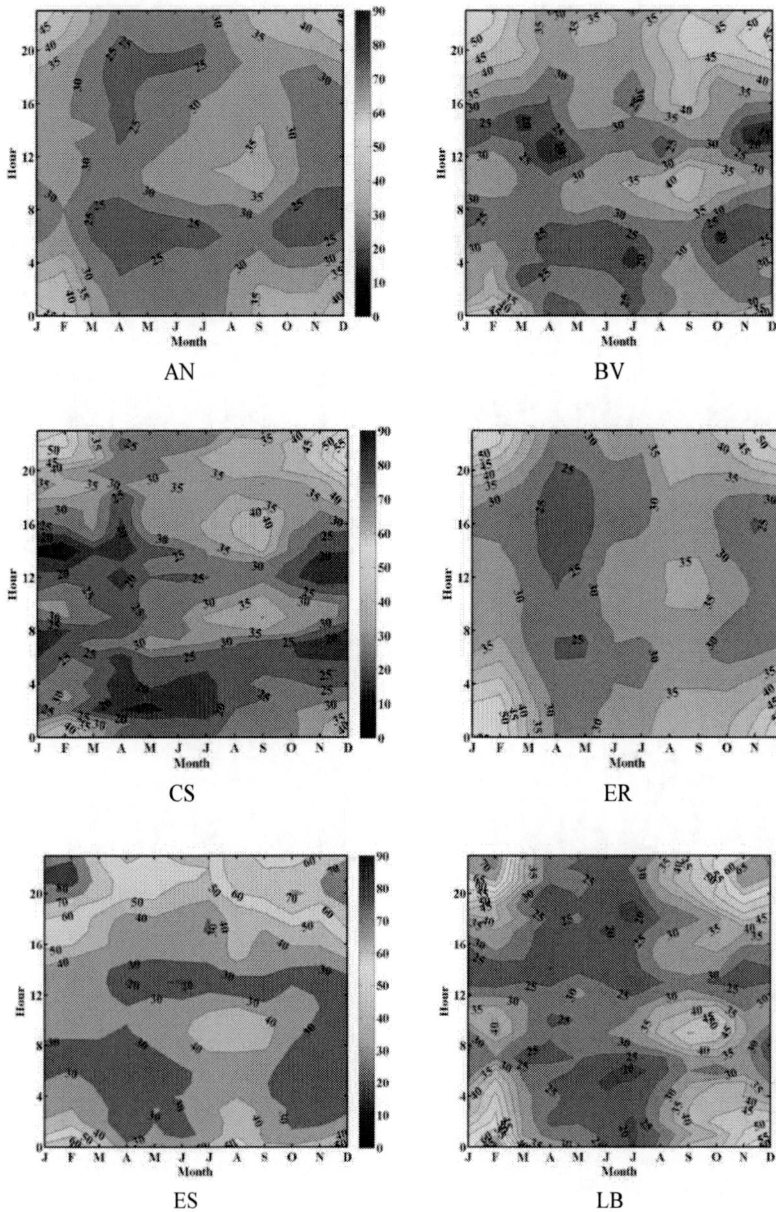


Figure 2. (Continued).

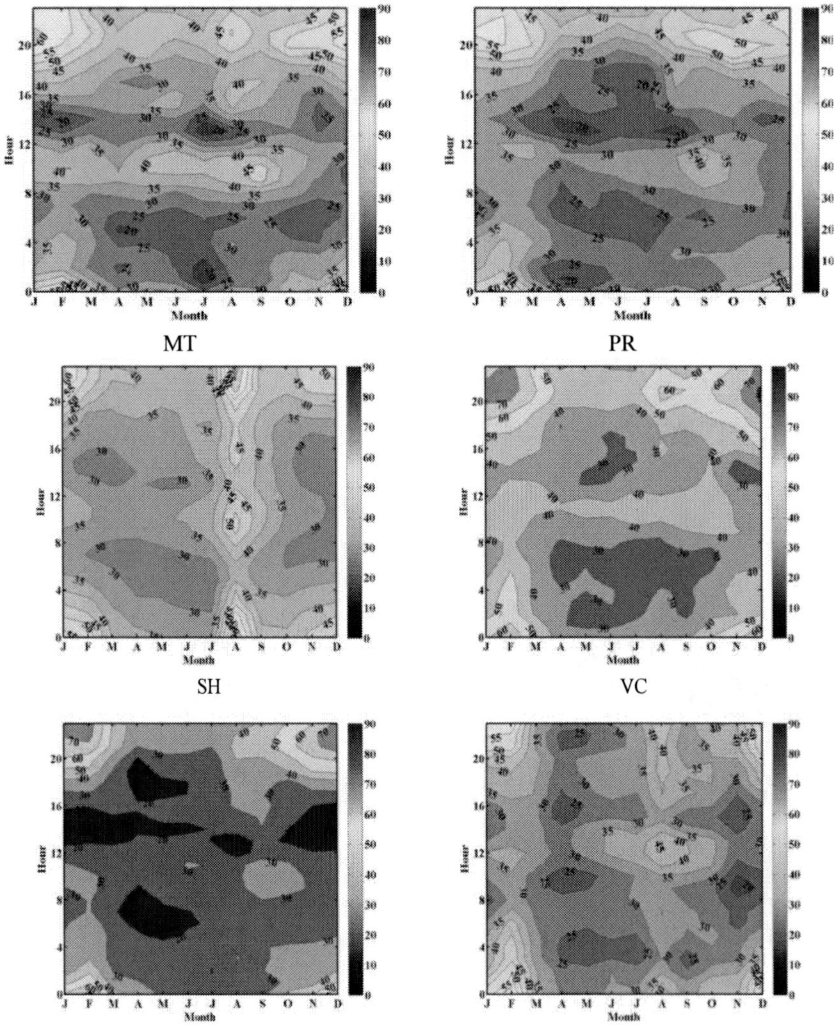


Figure 2. Annual average profiles of $PM_{10}$ concentrations (in μg $m^{-3}$) calculated for the studied period (2006-2010) at 12 monitoring sites.

# CONCLUSION

The data collected by the Air Quality Monitoring Network of Oporto Metropolitan Area showed that this region presents high concentrations of $PM_{10}$. The EU standard limits were surpassed several times (more than the allowed by the legislation) at some monitoring sites. However, $PM_{10}$ concentrations showed a decreasing tendency, presenting more acceptable values in the last three years of the analysed period. The application of PCA to $PM_{10}$ data confirmed few redundancies in the monitoring network. Only two monitoring sites often presented similar air pollution behaviours. Analysing the annual average profiles of $PM_{10}$ concentrations at all monitoring sites, high values were observed at night time during cold periods (January, February, November and December). This behaviour may be justified by the daily variation of the atmospheric boundary layer height that strongly influences the mixing of air pollutants.

## ACKNOWLEDGMENTS

J. C. M. Pires is grateful to *Comissão de Coordenação da Direcção Regional-Norte* for kindly providing the air quality data. J.C.M. Pires also acknowledges his Post-Doctoral fellowship (SFRH/BPD/66721/2009) supported by the Portuguese Foundation for Science and Technology (FCT), POPH-QREN and FSE.

## REFERENCES

[1] Brunekreef B., Holgate S. T. Air pollution and health. *Lancet* 2002; 360:1233-1242.

[2] Wenger D., Gerecke A. C., Heeb N. V., Hueglin C., Seiler C., Haag R., Naegeli H., Zenobi R. Aryl hydrocarbon receptor-mediated activity of atmospheric particulate matter from an urban and a rural site in switzerland. *Atmos Environ* 2009;43:3556-3562.

[3] Wilson J. G., Kingham S., Pearce J., Sturman A. P. A review of intraurban variations in particulate air pollution: Implications for epidemiological research. *Atmos Environ* 2005; 39:6444-6462.

[4] Yao Q., Li S. Q., Xu H. W., Zhuo J. K., Song Q. Studies on formation and control of combustion particulate matter in china: *A review. Energy* 2009;34:1296-1309.

[5] Ghafghazi S., Sowlati T., Sokhansanj S., Bi X., Melin S. Particulate matter emissions from combustion of wood in district heating applications. *Renew Sust Energ Rev* 2011;15:3019-3028.

[6] Lee R. G. M., Coleman P., Jones J. L., Jones K. C., Lohmann R. Emission factors and importance of pcdd/fs, pcbs, pcns, pahs and pm10 from the domestic burning of coal and wood in the uk. *Environ Sci Technol* 2005;39:1436-1447.

[7] Querol X., Viana M., Alastuey A., Amato F., Moreno T., Castillo S., Pey J., de la Rosa J., de la Campa A. S., Artinano B., Salvador P., dos Santos S. G., Fernandez-Patier R., Moreno-Grau S., Negral L., Minguillon M. C., Monfort E., Gil J. I., Inza A., Ortega L. A., Santamaria JM, Zabalza J. Source origin of trace elements in pm from regional background, urban and industrial sites of spain. *Atmos Environ* 2007;41:7219-7231.

[8] Gray H. A., Cass G. R. Source contributions to atmospheric fine carbon particle concentrations. *Atmos Environ* 1998;32:3805-3825.

[9] Monks P. S., Granier C., Fuzzi S., Stohl A., Williams M. L., Akimoto H., Amann M., Baklanov A., Baltensperger U., Bey I., Blake N., Blake RS, Carslaw K, Cooper OR, Dentener F, Fowler D, Fragkou E, Frost G. J., Generoso S., Ginoux P., Grewe V., Guenther A., Hansson H. C., Henne S., Hjorth J., Hofzumahaus A., Huntrieser H., Isaksen I. S. A., Jenkin M. E., Kaiser J., Kanakidou M., Klimont Z., Kulmala M., Laj P., Lawrence M. G., Lee J. D., Liousse C., Maione M., McFiggans G., Metzger A., Mieville A., Moussiopoulos N., Orlando J. J., O'Dowd C. D., Palmer P. I., Parrish D. D., Petzold A., Platt U., Poschl U., Prevot A. S. H., Reeves C. E., Reimann S., Rudich Y., Sellegri K., Steinbrecher R., Simpson D., ten Brink H., Theloke J., van der Werf G. R., Vautard R., Vestreng V., Vlachokostas C., von Glasow R. Atmospheric composition change - global and regional air quality. *Atmos Environ* 2009; 43:5268-5350.

[10] Koelemeijer R. B. A., Homan C. D., Matthijsen J. Comparison of spatial and temporal variations of aerosol optical thickness and particulate matter over europe. *Atmos Environ* 2006; 40:5304-5315.

[11] Pope C. A., Dockery D. W. Health effects of fine particulate air pollution: Lines that connect. *J Air Waste Manage* 2006;56:709-742.

[12] Harrison R. M., Yin J. X. Particulate matter in the atmosphere: Which particle properties are important for its effects on health? *Sci Total Environ* 2000;249:85-101.

[13] Thurston G. D., Ito K., Hayes C. G., Bates D. V., Lippmann M. Respiratory hospital admissions and summertime haze air-pollution in toronto, ontario - consideration of the role of acid aerosols. *Environ Res* 1994;65:271-290.

[14] Burnett R. T, Brook J, Dann T, Delocla C, Philips O, Cakmak S, Vincent R, Goldberg MS, Krewski D. Association between particulate- and gas-phase components of urban air pollution and daily mortality in eight canadian cities. *Inhal Toxicol* 2000;12:15-39.

[15] Kappos A. D., Bruckmann P., Eikmann T., Englert N., Heinrich U., Hoppe P., Koch E., Krause G. H. M., Kreyling W. G., Rauchfuss K., Rombout P., Schulz-Klemp V., Thiel W. R., Wichmann H. E. Health effects of particles in ambient air. *Int J Hyg Envir Heal* 2004; 207:399-407.

[16] Directive. Council directive 2008/50/ec, on ambient air quality and cleaner air for europe. *Official Journal of the European Communities* 2008; L151:1-44.

[17] Poschl U. Atmospheric aerosols: Composition, transformation, climate and health effects. *Angew Chem Int Edit* 2005;44:7520-7540.

[18] Lohmann U,, Feichter J. Global indirect aerosol effects: A review. *Atmos Chem Phys* 2005; 5:715-737.

[19] Haywood J., Boucher O. Estimates of the direct and indirect radiative forcing due to tropospheric aerosols: *A review. Rev Geophys* 2000;38:513-543.

[20] USCCSP. *Atmospheric aerosol properties and climate impacts*, Washington, DC: National Aeronautics and Space Administration; 2009.

[21] Ebi K. L., McGregor G. Climate change, tropospheric ozone and particulate matter, and health impacts. *Environ Health Persp* 2008; 116:1449-1455.

[22] Carvalho A., Monteiro A., Solman S., Miranda A. I., Borrego C. Climate-driven changes in air quality over europe by the end of the 21st century, with special reference to portugal. *Environ Sci Policy* 2010; 13:445-458.

[23] Jacob D. J., Winner D. A. Effect of climate change on air quality. *Atmos Environ* 2009; 43:51-63.

[24] Pires J. C. M., Pereira M. C., Alvim-Ferraz M. C. M., Martins F. G. Identification of redundant air quality measurements through the use of principal component analysis. *Atmos Environ* 2009;43:3837-3842.

[25] Pires J. C. M., Sousa S. I. V., Pereira M. C., Alvim-Ferraz M. C. M., Martins F. G. Management of air quality monitoring using principal component and cluster analysis - part i: So2 and pm10. *Atmos Environ* 2008; 42:1249-1260.

[26] Pires J. C. M., Sousa S. I. V., Pereira M. C., Alvim-Ferraz M. C. M., Martins F. G. Management of air quality monitoring using principal

component and cluster analysis - part ii: Co, no2 and o-3. *Atmos Environ* 2008; 42:1261-1274.

[27] Camdevyren H., Demyr N., Kanik A., Keskyn S. Use of principal component scores in multiple linear regression models for prediction of chlorophyll-a in reservoirs. *Ecol Model* 2005; 181:581-589.

[28] Harman H. H. *Modern factor analysis*. University of Chicago Press; 1976.

[29] Kannel P. R., Lee S., Kanel S. R., Khan S. P. Chemometric application in classification and assessment of monitoring locations of an urban river system. *Anal Chim Acta* 2007; 582:390-399.

[30] Mendiguchia C., Moreno C., Galindo-Riano M. D., Garcia-Vargas M. Using chemometric tools to assess anthropogenic effects in river water - a case study: Guadalquivir river (spain). *Anal Chim Acta* 2004; 515:143-149.

[31] Shrestha S., Kazama F. Assessment of surface water quality using multivariate statistical techniques: A case study of the fuji river basin, japan. *Environ Modell Softw* 2007; 22:464-475.

[32] Singh K. P., Malik A., Mohan D., Sinha S. Multivariate statistical techniques for the evaluation of spatial and temporal variations in water quality of gomti river (india) - a case study. *Water Res* 2004; 38:3980-3992.

[33] Singh K. P., Malik A., Sinha S. Water quality assessment and apportionment of pollution sources of gomti river (india) using multivariate statistical techniques - a case study. *Anal Chim Acta* 2005; 538:355-374.

[34] Gramsch E., Cereceda-Balic F., Oyola P., von Baer D. Examination of pollution trends in santiago de chile with cluster analysis of pm10 and ozone data. *Atmos Environ* 2006; 40:5464-5475.

[35] Pires J. C. M., Martins F. G., Alvim-Ferraz M. C. M., Pereira M. C. Evaluation of redundant measurements on the air quality monitoring network of lisbon and tagus valley. *Chemical Product and Process Modeling* 2009; 4:1-10.

[36] Pires J. C. M., Martins F. G. Evaluation of spatial variability of pm10 concentrations in london. *Water Air Soil Poll* 2012; 223:2287-2296.

[37] Ibarra-Berastegi G., Saenz J., Ezcurra A., Ganzedo U., de Argandona J. D., Errasti I., Fernandez-Ferrero A., Polanco-Martinez J. Assessing spatial variability of so2 field as detected by an air quality network using self-organizing maps, cluster, and principal component analysis. *Atmos Environ* 2009; 43:3829-3836.

[38] Afif C., Dutot A. L., Jambert C., Abboud M., Adjizian-Gérard J., Farah W., Perros P. E., Rizk T. Statistical approach for the characterization of no2 concentrations in beirut. *Air Quality, Atmosphere and Health* 2009; 2:57-67.

[39] Lau J., Hung W. T., Cheung C. S. Interpretation of air quality in relation to monitoring station's surroundings. *Atmos Environ* 2009; 43:769-777.

[40] Lu W. Z., He H. D., Dong L. Y. Performance assessment of air quality monitoring networks using principal component analysis and cluster analysis. *Build Environ* 2011; 46:577-583.

[41] Gao T., Xie S.-D., Bo Y., Zhao Y. Assessing pm10 and so2 networks using positive matrix factorization in beijing city. *Huanjing Kexue / Environmental Science* 2010; 31:560-565.

[42] Pereira M. C., Alvim-Ferraz M. C. M., Santos R. C. Relevant aspects of air quality in oporto (portugal): Pm10 and o-3. *Environ Monit Assess* 2005; 101:203-221.

[43] Yidana S. M., Ophori D., Banoeng-Yakubo B. A multivariate statistical analysis of surface water chemistry data - the ankobra basin, ghana. *J Environ Manage* 2008; 86:80-87.

[44] Sportisse B. *Fundamentals in air pollution: From processes to modelling*. Springer; 2010.

In: Atmosphere and Climate
Editors: A. Wright and S. Johnson

ISBN: 978-1-62417-433-9


*Chapter 4*

# DEVELOPMENT AND EVALUATION OF FIBROUS ADSORBENTS FOR CARBON DIOXIDE REMOVAL

***Xianfeng Wang[1,3], Yuhua Duan[1,2] and Bingyun Li[1,3,*]***

[1]National Energy Technology Laboratory-Regional University Alliance, (NETL-RUA) US

[2]National Energy Technology Laboratory, U.S. Department of Energy, Pittsburgh, PA, US

[3]Department of Orthopaedics, West Virginia University, Morgantown, WV, US

## ABSTRACT

The mitigation of carbon dioxide ($CO_2$) emissions has attracted worldwide attention because of the environmental and economic threats posed by possible global climate changes. Considering that the utilization of fossil energy carriers will continue and may even increase over the coming decades, advanced $CO_2$ capture materials and technologies are of the utmost importance. Micro/nanofibers have unique porous structure, high surface area, and good mechanical properties and thereby are

* Correspondence to: Bingyun Li, PhD, Associate Professor Department of Orthopaedics, West Virginia University Morgantown, WV 26506-9196, USA Tel: 1-304-293-1075, Fax: 1-304-293-7070, Email: bili@hsc.wvu.edu (B. Li) URL: http://www.hsc.wvu.edu/som/ortho/nanomedica-group/.

attractive candidates as advanced sorbents for $CO_2$ removal. Fibrous sorbents may have high $CO_2$ capture capacity and little resistance to bulk gas flow and thereby have fast kinetics. In this chapter, we have summarized the recent progress in the development of fibrous sorbents (e.g. amine-modified fibers, carbon fibers, hollow fiber membranes, etc.) for $CO_2$ removal, described the types of fiber materials that have been studied, and discussed their fabrication variables and $CO_2$ adsorption performance in details. This review may pave the way for developing advanced fibrous adsorbents for $CO_2$ removal from power plants and even from the atmosphere.

**Keywords:** $CO_2$ removal, fibrous adsorbent, fiber, nanofiber, membrane

# 1. INTRODUCTION

Global carbon dioxide ($CO_2$) emission caused by escalating energy use may have resulted in a series of environmental problems such as anomalous climate change and rising of sea levels [1-3]. Studies have revealed that the concentration of $CO_2$ in the atmosphere has increased by ~ 23% to a value of 380 ppm during the last half century and it is projected to continue to rise if anthropogenic sources remain unchecked [4]. In the U.S., over 94% of the $CO_2$ emission comes from the combustion of carbon-based fossil fuels and thus the U.S. Department of Energy (DOE) issued a carbon sequestration roadmap in 2009 aiming to achieve 90% $CO_2$ capture at an increased cost of electricity of no more than 35% for the post-combustion process by 2020 [4-6]. Technologies for $CO_2$ capture and storage (CCS) present some of the most promising and effective options for large-scale reductions in $CO_2$ emissions [1]. As a means to remove $CO_2$ from flue gases on a large-scale, there are four main methods to separate $CO_2$ from other gases: solid sorbents, cryogenic distillation, solvent, and membrane [7]. It is evident that the progress and achievement of most of these technologies depends heavily on the development of materials.

Carbon fibers and their composites as well as hollow fiber membranes have been developed for gas separation [8, 9]. Benefiting from the unique properties (e.g. high porosity, high surface area, and good mechanical properties) of fibrous structures [10, 11], fibrous adsorbent has emerged in the past decade as a new class of adsorption materials; considerable efforts have been applied on their adsorbent design and optimization [12-14]. In this chapter, the recent progress of fibrous sorbents used for $CO_2$ capture and separation has been reviewed.

# 2. Fibrous Adsorbents for $CO_2$ Capture

## 2.1. Amine-modified Fibrous Adsorbents

Recently, modification with amine groups on the surfaces of adsorption materials has attracted much attention due to the potential low energy requirement, low cost, and ease of applicability over a wide range of temperatures and pressures [15]. Up to now, amine-modified particle adsorbents such as amine-modified MCM-41, SBA-15, polymethyl-methacrylate, fly ash, zeolite, and activated carbon have been extensively investigated for $CO_2$ capture [16-18]. All of the supports mentioned above are microporous or mesoporous particles with a high surface area and a large pore volume. The supports can be amine-treated by coating their surfaces with amine-rich polymers or compounds, for example, polyethylenimine (PEI), allylamine, and aminosilanes [19]. In this section, we have focused on the recent progress of some amine-modified fibrous adsorbents.

### *2.1.1. Glass Fibrous Adsorbent*

In 1991, Andreopoulos and coworkers [20] demonstrated that fibrous adsorbents based on PEI coated polypropylene fibers had high adsorption capacities to acidic gases (e.g. $SO_x$). More recently, glass fiber has received much attention due to its high surface area, low price, and convenience to be used as a matrix (Figure 1a). Inspired by Andreopoulos' work, Li et al. [21] developed a novel fibrous adsorbent for $CO_2$ capture by coating PEI on a glass fiber matrix using epoxy resin as a crosslinking agent. They systematically investigated the fabrication variables (e.g. crosslinking agent dosage, coating weight, moisture, adsorption temperature, and $CO_2$ concentration of the simulated flue gas) that may affect the adsorption capacity of their fibrous adsorbents and found that a maximum adsorption capacity of 276.96 mg of $CO_2$/(g of PEI) was obtained at a PEI/epoxy resin ratio of 10:1. Moisture was observed to enhance the adsorption of $CO_2$; the $CO_2$ adsorption capacity of the fibrous adsorbent in the presence of moisture could be 19 times higher than that in the absence of moisture. It is worthy noting that higher molecular weight and larger steric obstruction of the cross- linking agent used (i.e. epoxy resin) led to poorer diffusion of $CO_2$ in the fibrous adsorbent and thus poorer kinetics. Therefore, it is of interest to identify better cross-linking agents with lower molecular weight to replace the epoxy resin in order to achieve better performance. Li et al. replaced epoxy resin (MW=370) with epichlorohydrin (ECH, MW=92.5) and developed adsorbents of PEI modified glass fibers [23]. As expected, the PEI modified glass fibers improved adsorption performance (e.g. higher $CO_2$ capacity, faster kinetics, and better regenerability) [22]. As shown in Figure 1b, PEI was deposited as a porous layer on the surfaces of glass fibers, and this process might have contributed to the more efficient contact with $CO_2$ thereby leading to a higher adsorption rate and capacity.

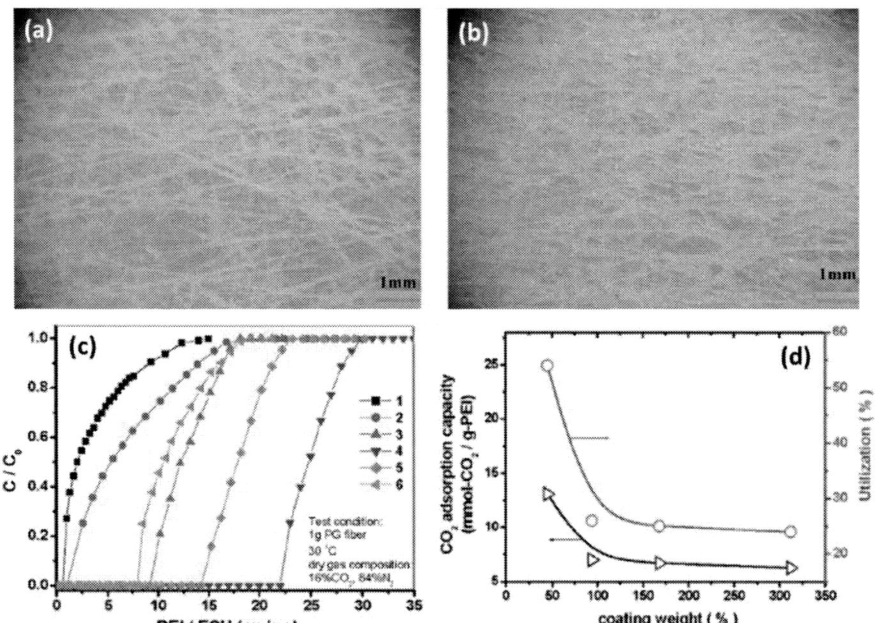




Figure 1. Micrograph images of (a) glass fiber and (b) PEI modified glass fibers; (c) Breakthrough curves of $CO_2$ adsorption on PEI modified glass fibers with different PEI/ECH ratios: (1) 5:1; (2) 10:1; (3) 15:1; (4) 20:1; (5) 25:1; and (6) 30:1. (d) $CO_2$ adsorption capacity *vs.* coating weight of PEI modified glass fibers.

The $CO_2$ adsorption capacities of the PEI modified glass fiber adsorbents with different coating weights are shown in Figure 1c. The maximum $CO_2$ adsorption capacity of 13.08 mmol $CO_2$/(g of PEI) was achieved at the lowest coating weight of 45 wt.%. The adsorption rate was rapid in the beginning phase, leading to a nearly zero concentration of $CO_2$ in effluent gas for several minutes. It is obvious that the $CO_2$ adsorption capacity does not increase with increasing coating weight (Figure 1d).

### *2.1.2. Fibrous Polyacrylonitrile (PAN) Adsorbent*

Polymer supports for amines have attracted considerable attention since they are light in weight, flexible, and easy to handle [7]. Solid amine adsorbents using synthetic polymer fibers as matrix may offer advantages for $CO_2$ adsorption because of the high external surface area, potential low pressure drops, and flexibility of the matrix fibers [23].

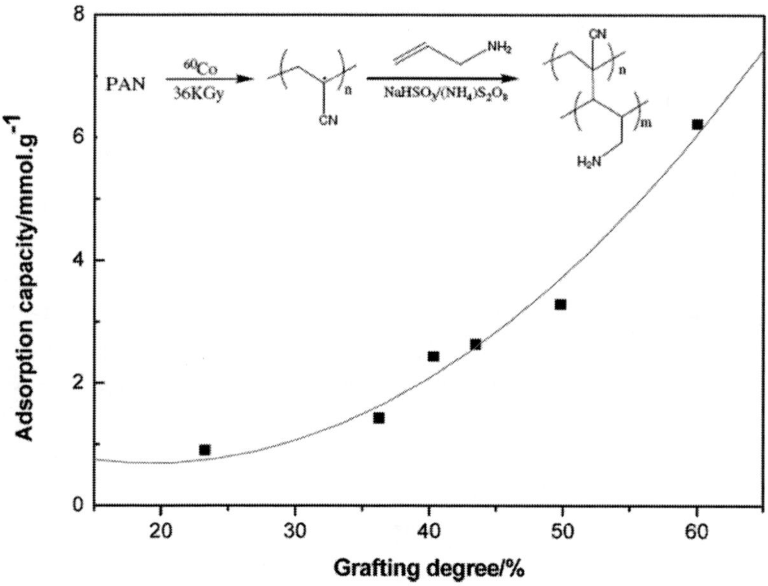


Reprinted (adapted) with permission from [19]. Copyright (2010) American Chemical Society.

Figure 2. Effect of grafting degree on adsorption capacity (adsorption temperature: 22 °C; concentration of $CO_2$: 15.2%). Inset shows the preparation procedure of irradiation grafting of allylamine onto PAN fiber.

For amine-impregnated sorbents, $CO_2$ molecules could easily be adsorbed on the surface of the amines especially at low temperature, but the diffusion of $CO_2$ molecules into the interior of the amines is difficult [13]. Additionally, poor adhesion between amine and the support would lead to easy leakage from the support thus spoiling the regeneration performance. To enhance the adhesion between amines and the support, the grafting copolymerization method was introduced. For example, Yang et al. [19] demonstrated that solid amine-containing fibrous adsorbent could be prepared by pre-irradiation grafting copolymerization of allylamine onto PAN fiber (PAN-AF), using the redox system of $(NH_4)_2S_2O_8/NaHSO_3$ as an initiator (Figure 2, inset). They found that the reaction conditions such as reaction time, temperature, monomer concentration, and amount of the initiator had great effects on $CO_2$ adsorption performance. As presented in Figure 2, higher grafting degree resulted in higher $CO_2$ uptake and the adsorption capacity of PAN-AF reached 6.22 mmol $CO_2$/(g of PAN-AF) at the grafting degree of 60.0 wt.%. They attributed the good performance of this fibrous adsorbent to the fibrous structure, which might have reduced the resistance to gas flow thereby maximizing sorbent performance while minimizing energy consumption, and the highly stable interface (i.e. grafting copolymerization of amines onto fibers) between amines and the support [19].

### *2.1.3. Nanofibrillated Cellulose (NFC) Adsorbent*

NFC is an abundant natural material composed of cellulose fibril aggregates (10~100 nm in diameter and several micrometers in length) [14]. NFC is rich in surface hydroxyl groups, which can be used to covalently anchor chemical functionalities to its cellulose backbone [24, 25]. Amine functionality has been introduced on NFC through grafting of aminosilanes [26]. However, the hydrolysis and self-condensation of aminosilanes with three alkoxy groups may lead to the build-up of three-dimensional siloxane networks, which is unfavorable for $CO_2$ capture; the siloxane networks may block the functional amine sites and the pores in the support [27].

To avoid the undesired blocking of functional amine sites, Gebald and coworkers [12] developed a new adsorbent based on aminosilanes with two alkoxy functions (N-(2-aminoethyl)-3 aminopropylmethyldimethoxysilane or AEAPDMS. The adsorbent was synthesized through freeze-drying (FD) of an aqueous suspension of NFC and AEAPDMS. Before FD, the NFC hydrogel morphology was composed of entangled cellulose nanofibrils of submicrometer diameter (Figure 3a). After FD, the NFC-FD morphology exhibited cellulose sheet structures with single irregularly distributed cellulose nanofibrils attached to the cellulose sheets (Figure 3b). The presence of AEAPDMS in the NFC suspension (i.e. AEAPDMS-NFC-FD) further increased the sheet-forming tendency and eliminated single cellulose nanofibrils (Figure 3c). The adsorption rate of AEAPDMS-NFC-FD was comparable to those of hyperbranched aminosilica, PEI/silica, 3-aminopropyltrimethoxysilane modified PEI/silica (A-PEI/silica), and tetraethyl orthotitanate modified PEI/silica (T-PEI/silica) in their powder forms (Figure 3d). The favorable $CO_2$ uptake rate could be explained by the highly porous structure of AEAPDMS-NFC-FD, which is composed of cellulose sheets separated by pores in the micrometer range. During 2 h of $CO_2$ adsorption and 1 h of $CO_2$ desorption in Ar, the average cyclic $CO_2$ capacity was 0.695 mmol $CO_2$/g. The AEAPDMS-NFC-FD adsorbent was found to be stable for at least 20 adsorption/desorption cycles (Figure 3e) [12].

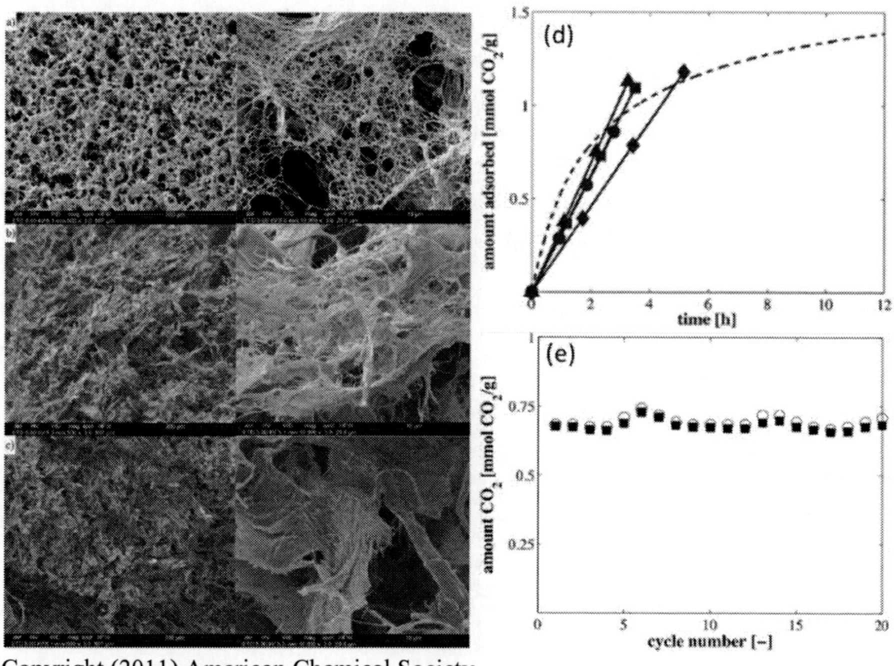




Figure 3. (a) Scanning electron microscopy images of NFC hydrogel, (b) NFC-FD, (c) AEAPDMS-NFC-FD. (d) Comparison of $CO_2$ capacity as a function of time of AEAPDMS-NFC-FD (dashed line), hyperbranched aminosilica (spheres), PEI/silica (rhombuses), A-PEI/silica (triangles), and T-PEI/silica (squares). (e) Multicycle adsorption (squares) and desorption (spheres) experiments with AEAPDMS-NFC-FD. Reprinted (adapted) with permission from [12].

### *2.1.4. Mesoporous Silica/Cotton Fibrous Adsorbent*

Ordered mesoporous silica materials have evoked much interest in the past ten years due to their potential applications in catalysis, sorbents, drug delivery, optical devices, and nanostructure templates [28, 29]. Recently, one interesting research orientation is the utilization of the as-synthesized mesoporous silica as a crucial template component for functional materials [30]. The template is utilized to disperse amines, and thus enhances the adsorption capacity of this system by modifying the interaction between $CO_2$ and the amines. However, the industrial application of these mesoporous materials remains a challenge, i.e. fine powder can be dissipated with carrier gas due to the small size of the particles. Recently, Ma et al. [31] developed a method to shape the mesoporous silica (i.e. SBA-3) by *in situ* synthesis on the surfaces of cotton fibers. These shaped composite materials were subsequently modified with tetraethylenepentamine (TEPA) to form composite fibrous adsorbents, as shown in Figure 4. During the synthesis of mesoporous silica, the strong bibulous ability of cotton fiber enables it to absorb the whole reaction solution. Cross-linked cotton fibers were used to construct a flexible three-dimensional (3D) netlike matrix to disperse and support mesoporous silica particles thus preventing over-packing of the particles. The reported SBA-3/cotton (S/Cot) fiber composite materials exhibited considerable $CO_2$ adsorption capacity [106 mg/(g of S/Cot composites)] and stable cyclic adsorption/desorption performances. This approach may produce inexpensive shaped composite materials with high-performance, and the composite fibrous adsorbents may have advantages over $CO_2$ powder adsorbents.

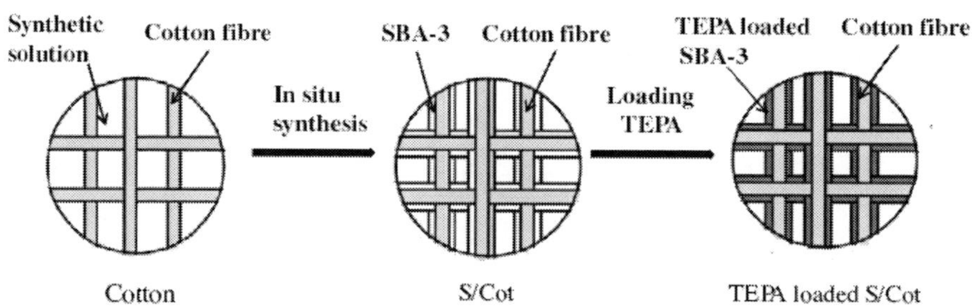




Figure 4. Schematic diagrams for the formation of S/Cot and TEPA loaded S/Cot fibrous adsorbents. Reprinted with permission from [31].

## 2.2. Carbon Fiber Adsorbents

Porous carbon materials are well-known $CO_2$ adsorbents due to their highly developed porosity, extended surface area, flexible surface chemistry, and high thermal stability. Recently, some new classes of carbon materials (e.g. carbon fiber, carbon molecular sieves) have emerged as adsorbents for gas separation and storage. Carbon fiber refers to fibers which are at least 90 wt.% carbon in composition obtained by the controlled pyrolysis of an appropriate precursor material (e.g. pitch, PAN, rayon, nonheterocyclic aromatic polymers) [32]. The structures of fibers can be crystalline or amorphous, and the crystalline form of carbon is graphite. When the proportion of graphite is high, the fiber is called graphite fiber.

#### *2.2.1. Carbon Fiber Composite Molecular Sieve (CFCMS)*

The CFCMS is a monolithic activated carbon that is composed of petroleum pitch-derived carbon fiber and a phenolic resin-derived binder. The U.S. Oak Ridge National Laboratory has investigated CFCMS applications in a variety of gas separation and gas storage [33]. Activation of the CFCMS creates microporosity in the carbon fibers, yielding high micropore volumes (>0.5 $cm^3/g$) and BET surface areas (>1000 $m^2/g$). Moreover, the CFCMS material is a rigid, strong monolith with an open structure that allows the free-flow of fluid/gas through the material. This combination of properties provides an adsorbent material that has several distinct advantages over granular adsorbents in gas separation systems such as pressure swing adsorption units. The room temperature $CO_2$ adsorption capacity of CFCMS is about 120 mg/(g of CFCMS), and the CFCMS exhibited rapid adsorption and desorption compared to zeolites and conventional granular activated carbons [33].

#### *2.2.2. PAN-based Activated Carbon Fiber (PAN-ACF)*

PAN-ACFs with high surface areas and nitrogen contents can be made from PAN fibers through a partial gasification reaction. Much research has been devoted to the preparation of PAN-ACFs from modified PAN, and some attention has been paid to the use of PAN-ACFs for $CO_2$ removal. Hierarchical porous PAN-ACFs with a large BET surface area (2231 $m^2/g$) and a high pore volume (1.16 $cm^3/g$) were made from PAN fibers through pre-oxidation and chemical activation. This type of material contains a large number of nitrogen-containing groups (N content > 8.1 wt.%) and thereby basic sites, leading to a faster adsorption rate and a higher $CO_2$ adsorption capacity (2.4 mmol/g). Moreover, PAN-ACF adsorbents had stable $CO_2$ adsorption/desorption performance under multiple cycle conditions [34].

#### *2.2.3. Graphite Nanofiber (GNF)*

As one of the most important carbon materials, graphite fibers have great potential for applications in gas storage, electrodes, filler, and catalyst supports [35-37]. However, graphite fibers have relatively lower specific surface area compared to other carbon materials, and effort therefore has diverted into engineering graphite fibers into nanosize or porous structures in order to increase their surface area and ultimately enhance their $CO_2$ adsorption capacity. For instance, Meng et al. [35] found that porous GNFs (Figure 5) could be fabricated using a KOH etching method at temperatures

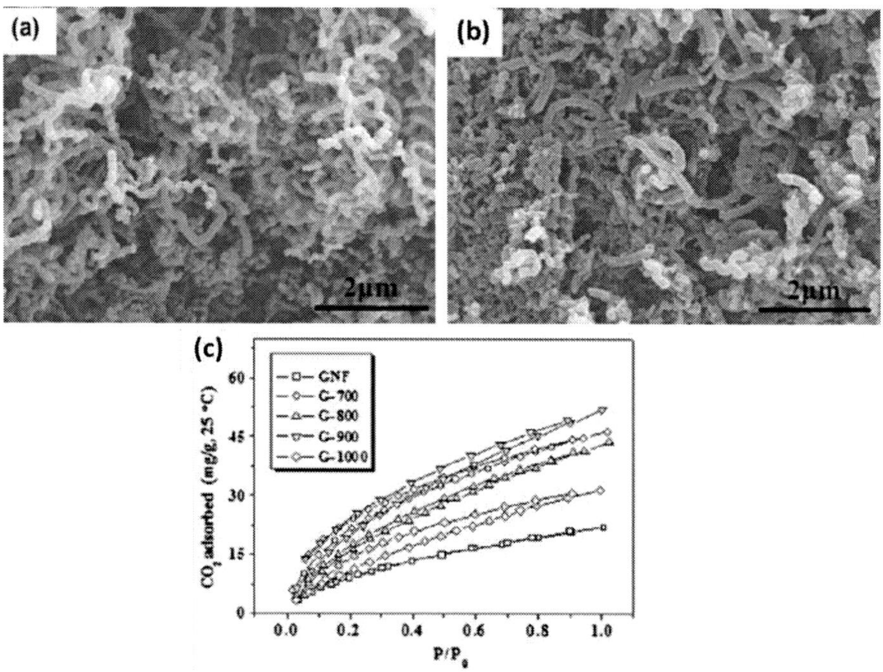




Figure 5. Typical field emission scanning electron microscopy images (a) pristine and (b) chemical-treated GNFs. (c) $CO_2$ isotherms of the pristine and chemical-treated GNFs [35]. Reprinted with permission from [35].

in the 700-1000°C range. Chemical activation of GNFs (Figure 5) with KOH begins with a chemical reduction of the hydroxide, followed by intercalation of the metal into the graphene layers, leading to increased microporosity. Typical adsorption/desorption isotherms of $N_2$ on the pristine and chemical-treated GNFs showed that KOH activation promoted the development of pores in the samples. The $CO_2$ adsorption isotherms revealed that GNFs treated under 900°C had the highest BET surface area (567 $m^2/g$) and the best $CO_2$ adsorption capacity (59.2 mg/g) (Figure 5c).

## 2.3. Hollow Fiber Membranes for $CO_2$ Removal

Membranes have been studied to separate $CO_2$ from flue gas [38-42]. Benefiting from the high ratio of membrane area per volume, hollow fiber membrane geometry is expected to have the potential to overcome the disadvantages (e.g. phase dispersion, limited mass transfer areas) of conventional membranes and lead $CO_2$ capture technology to a new level when integrated with chemical absorption [43]. The compact modular structure of membrane contactors also provides much larger gas-liquid interfaces and the flexibility to scale up or down. Up to now, hollow fiber membranes have been constructed using a number of different polymeric materials, including polysulfones, polyethersulfones, polyphenylene oxide, polyetherimide, and polyimides [44-46]. Because of the potential advantages offered by hollow fiber membranes, considerable academic research has been conducted [47, 48]. For instance, Karoor et al. [47] carried out comprehensive experiments and simulated the gas-liquid absorption in a microporous hydrophobic hollow fiber device using $CO_2$, $SO_2$, $CO_2$-$N_2$, and $SO_2$-air mixtures as feed gases and water as an absorbent. To solve the gas absorption performance deterioration of membranes caused by membrane wetting, Lv et al. [39] developed superhydrophobic polypropylene hollow fiber membranes by depositing a rough hydrophobic layer on membrane surfaces. The membrane-absorbent interaction results demonstrated that the modification treatment effectively enhanced the stability and maintained the superhydrophobicity of fibers contacting with the absorbent.

# PERSPECTIVE AND CONCLUSION

$CO_2$ removal has been attracting broad attention from both science and technology. In this chapter, we have reviewed the recent progress in developing fibrous sorbents (e.g. amine-modified fibers, carbon fibers, graphite nanofibers, hollow fibers, etc.) with potential applications for $CO_2$ removal. Clearly, the selection of capture materials is essential for any technologies in $CO_2$ removal. In general, fibrous materials have advantages such as ease of design and synthesis, high porosity, tailored pore properties, high surface area, and good mechanical properties. These properties make them highly attractive for the development of adsorbents with high $CO_2$ capture capacity and fast kinetics. However, further exploration is needed on several important issues, which may hamper their practical application, as discussed below.

The gas flows treated in practical $CO_2$ removal always contain water and it is not economically feasible to dry the flue gas by an additional process before separation. Therefore, adsorption materials require a high tolerance to water or to be superhydrophobic. However, only a few evaluations of the effect of water on the separation performances of fibrous adsorbents have been reported. Addressing this issue should include research on both the physical co-adsorption of water in the fibrous materials and the material structure and functional design.

Some fibrous materials do not have $CO_2$ adsorption ability until they are modified with amine groups as described in this chapter. However, the undesired blocking of fiber surfaces with thick amines will lower the $CO_2$ capture performance of the adsorbents. Further research should focus on the achievement of effective surface modification strategies. For example, surface modifying fibrous materials using layer-by-layer nano-assembly technology [49, 50] via depositing $CO_2$-adsorbing amine polymer would be a potential method to solve this problem.

As demonstrated in this chapter, nanostructured fibrous adsorbents (e.g. NFC, GNF) exhibited improved $CO_2$ adsorption performance due to their large specific surface areas. Therefore, the design of nanofibrous adsorbents for $CO_2$ capture and the exploitation of new nanofiber fabrication methods (e.g. electrospinning [10]) will be of interest. Additionally, paralleling experimental studies and molecular simulation of fibrous adsorbents must be further developed.

Despite the numerous challenges surrounding $CO_2$ removal, the ease and low cost of the synthesis of fibrous adsorbents in large scale will be extremely important in determining their potential for industrial applications.

## ACKNOWLEDGMENTS

As part of the National Energy Technology Laboratory's Regional University Alliance (NETL-RUA), a collaborative initiative of the NETL, this technical effort was performed under the RES contract DE-FE0004000. Support from WV NASA EPSCoR was also acknowledged. Any opinions, findings, conclusions or recommendations expressed in this material are those of the author(s) and do not necessarily reflect the views of the funding agencies or their institutions. The authors thank Suzanne Smith for proofreading.

## Disclaimer

This project was funded by the Department of Energy, National Energy Technology Laboratory, an agency of the United States Government, through a support contract with URS Energy andConstruction, Inc. Neither the United States Government nor any agency thereof, nor any of their employees, nor URS Energy and Construction, Inc., nor any of their employees, makes any warranty, expressed or implied, or assumes any legal liability or responsibility for the accuracy, completeness, or usefulness of any information, apparatus, product, or process disclosed, or represents that its use would not infringe privately owned rights. Reference herein to any specific commercial product, process, or service by trade name, trademark, manufacturer, or otherwise, does not necessarily constitute or imply its endorsement, recommendation, or favoring by the United States Government or any agency thereof. The views and opinions of authors expressed herein do not necessarily state or reflect those of the United States Government or any agency thereof.

## Conflict of Interest

None.

## References

[1] N. Du, H.B. Park, M.M. Dal-Cin, M.D. Guiver, *Energy and Environmental Science*, 5 (2012) 7306-7322.

[2] W. Xing, C. Liu, Z. Zhou, L. Zhang, J. Zhou, S. Zhuo, Z. Yan, H. Gao, G. Wang, S.Z. Qiao, *Energy and Environmental Science*, 5 (2012) 7323-7327.

[3] R. Banerjee, A. Phan, B. Wang, C. Knobler, H. Furukawa, M. O'Keeffe, O.M. Yaghi, *Science*, 319 (2008) 939-943.

[4] J. Liu, P.K. Thallapally, B.P. McGrail, D.R. Brown, *Chemical Society Reviews*, 41 (2012) 2308-2322.

[5] J.P. Ciferno, T.E. Fout, A.P. Jones, J.T. Murphy, *Chemical Engineering Progress*, 105 (2009) 33-41.
[6] P.H. Stauffer, G.N. Keating, R.S. Middleton, H.S. Viswanathan, K.A. Berchtold, R.P. Singh, R.J. Pawar, A. Mancino, *Environmental Science and Technology*, 45 (2011) 8597-8604.
[7] S. Choi, J.H. Drese, C.W. Jones, *ChemSusChem*, 2 (2009) 796-854.
[8] W. Yave, A. Car, K.V. Peinemann, *Journal of Membrane Science*, 350 (2010) 124-129.
[9] P. Luis, T. Van Gerven, B. Van der Bruggen, *Progress in Energy and Combustion Science*, 38 (2012) 419-448.
[10] X. Wang, B. Ding, J. Yu, M. Wang, *Nano Today*, 6 (2011) 510-530.
[11] D. Li, Y. Xia, *Advanced Materials*, 16 (2004) 1151-1170.
[12] C. Gebald, J.A. Wurzbacher, P. Tingaut, T. Zimmermann, A. Steinfeld, *Environmental Science and Technology*, 45 (2011) 9101-9108.
[13] X. Wang, V. Schwartz, J.C. Clark, X. Ma, S.H. Overbury, X. Xu, C. Song, *The Journal of Physical Chemistry C*, 113 (2009) 7260-7268.
[14] D. Klemm, B. Heublein, H.P. Fink, A. Bohn, *Angewandte Chemie International Edition*, 44 (2005) 3358-3393.
[15] T. Filburn, J. Helble, R. Weiss, *Industrial and Engineering Chemistry Research*, 44 (2005) 1542-1546.
[16] X. Xu, C. Song, B.G. Miller, A.W. Scaroni, *Fuel Processing Technology*, 86 (2005) 1457-1472.
[17] C. Knöfel, J. Descarpentries, A. Benzaouia, V. Zeleňák, S. Mornet, P. Llewellyn, V. Hornebecq, *Microporous and Mesoporous Materials*, 99 (2007) 79-85.
[18] M. Plaza, C. Pevida, A. Arenillas, F. Rubiera, J. Pis, *Fuel*, 86 (2007) 2204-2212.
[19] Y. Yang, H. Li, S. Chen, Y. Zhao, Q. Li, *Langmuir*, 26 (2010) 13897-13902.
[20] A. Andreopoulos, J. Economy, *Polymers for Advanced Technologies*, 2 (1991) 87-91.
[21] P. Li, S. Zhang, S. Chen, Q. Zhang, J. Pan, B. Ge, *Journal of Applied Polymer Science*, 108 (2008) 3851-3858.
[22] P. Li, B. Ge, S. Zhang, S. Chen, Q. Zhang, Y. Zhao, *Langmuir*, 24 (2008) 6567-6574.
[23] M. Sevilla, P. Valle-Vigón, A.B. Fuertes, *Advanced Functional Materials*, 21 (2011) 2781-2787.

[24] S. Eichhorn, A. Dufresne, M. Aranguren, N. Marcovich, J. Capadona, S. Rowan, C. Weder, W. Thielemans, M. Roman, S. Renneckar, *Journal of Materials Science*, 45 (2010) 1-33.

[25] M.A. Hubbe, O.J. Rojas, L.A. Lucia, M. Sain, *BioResources*, 3 (2008) 929-980.

[26] J. Lu, P. Askeland, L.T. Drzal, *Polymer*, 49 (2008) 1285-1296.

[27] F. Beari, M. Brand, P. Jenkner, R. Lehnert, H.J. Metternich, J. Monkiewicz, H. Siesler, *Journal of Organometallic Chemistry*, 625 (2001) 208-216.

[28] F. Hoffmann, M. Cornelius, J. Morell, M. Fröba, *Angewandte Chemie International Edition*, 45 (2006) 3216-3251.

[29] M.E. Davis, *Nature*, 417 (2002) 813-821.

[30] L. Ma, K. Han, X. Ding, Y. Chun, J. Zhu, *Journal of Nanoscience and Nanotechnology*, 11 (2011) 4079-4088.

[31] K.K. Han, L. Ma, H.M. Zhao, X. Li, Y. Chun, J.H. Zhu, *Microporous and Mesoporous Materials*, 151 (2012) 157-162.

[32] R. Thiruvenkatachari, S. Su, H. An, X.X. Yu, *Progress in Energy and Combustion Science*, 35 (2009) 438-455.

[33] T.D. Burchell, R.R. Judkins, *Energy Conversion and Management*, 37 (1996) 947-954.

[34] W. Shen, S. Zhang, Y. He, J. Li, W. Fan, *Journal of Materials Chemistry*, 21 (2011) 14036-14040.

[35] L.Y. Meng, S.J. Park, *Journal of Colloid and Interface Science*, 352 (2010) 498-503.

[36] G. Chandrasekar, W.J. Son, W.S. Ahn, *Journal of Porous Materials*, 16 (2009) 545-551.

[37] B.J. Kim, Y.S. Lee, S.J. Park, *Journal of Colloid and Interface Science*, 318 (2008) 530-533.

[38] R.P. Lively, D.P. Leta, B.A. DeRites, R.R. Chance, W.J. Koros, *Chemical Engineering Journal*, 171 (2011) 801-810.

[39] Y. Lv, X. Yu, J. Jia, S.T. Tu, J. Yan, E. Dahlquist, *Applied Energy*, 90 (2012) 167-174.

[40] M.K. Barillas, R.M. Enick, M. O'Brien, R. Perry, D.R. Luebke, B.D. Morreale, *Journal of Membrane Science*, 372 (2011) 29-39.

[41] J. Ilconich, C. Myers, H. Pennline, D. Luebke, *Journal of Membrane Science*, 298 (2007) 41-47.

[42] C. Myers, H. Pennline, D. Luebke, J. Ilconich, J.K. Dixon, E.J. Maginn, J.F. Brennecke, *Journal of Membrane Science*, 322 (2008) 28-31.

[43] R. Wang, D.F. Li, D.T. Liang, *Chemical Engineering and Processing: Process Intensification*, 43 (2004) 849-856.
[44] R. Wang, H. Zhang, P. Feron, D. Liang, *Separation and Purification Technology*, 46 (2005) 33-40.
[45] H.Y. Zhang, R. Wang, D.T. Liang, J.H. Tay, *Journal of Membrane Science*, 279 (2006) 301-310.
[46] M. Sandru, S.H. Haukebų, M.B. Hägg, *Journal of Membrane Science*, 346 (2010) 172-186.
[47] S. Karoor, K.K. Sirkar, *Industrial and Engineering Chemistry Research*, 32 (1993) 674-684.
[48] S. Koonaphapdeelert, Z. Wu, K. Li, *Chemical Engineering Science*, 64 (2009) 1-8.
[49] B. Jiang, V. Kish, D.J. Fauth, M.L. Gray, H.W. Pennline, B. Li, *International Journal of Greenhouse Gas Control*, 5 (2011) 1170-1175.
[50] B. Li, B. Jiang, D.J. Fauth, M.M.L. Gray, H.W. Pennline, G.A. Richards, *Chemical Communications*, 47 (2011) 1719-1721.

# INDEX

## D



## E

## J



## K



## L



## M



## N